SOLAR ENERGY On A SHOESTRING BUDGET

Includes BONUS READING from

HOW TO BUY A CAR
Without Losing Your Shirt

—and—

PHOTOGRAPHIC COMPOSITION

© 2018

**This book copyright 2018 by
Crossroads Publishing of Florida
P. O. Box 222
Worthington Springs, FL 32697
www.cpubfl.com
All rights reserved**

This document is geared toward providing exact and reliable information regarding the topic and issue covered. The publication is sold with the idea that the publisher is not required to render legal, accounting, officially permitted, or otherwise qualified services. If advice is necessary, legal or professional, a practiced individual in the profession should be ordered.

- From a Declaration of Principles which was accepted and approved equally by a Committee of the American Bar Association and a Committee of Publishers and Associations.

In no way is it legal to reproduce, duplicate, or transmit any part of this document in either electronic means or in printed format. Recording of this publication is strictly prohibited and any storage of this document is not allowed unless with written permission from the publisher. All rights reserved.

The information provided herein is consistent, in that any liability, in terms of inattention or otherwise, by any usage or abuse of any policies, processes, or directions contained within is the solitary and utter responsibility of the recipient reader. Under no circumstances will any legal responsibility or blame be held against the publisher or author for any reparation, damages, or monetary loss due to the information herein, either directly or indirectly Respective authors own all copyrights not held by the publisher.

The information herein is offered for informational purposes solely, and is universal as so. The presentation of the information is without contract or any type of guarantee assurance.

The trademarks that are used are without any consent, and the publication of the trademark is without permission or backing by the trademark owner. All trademarks and brands within this book are for clarifying purposes only and are the owned by the owners themselves, not affiliated with this document.

• TABLE OF CONTENTS

Copyright Notice..5

Chapter 1: General Discussion..7

Chapter 2: What you Need and What you need to Know about it..17

Chapter 3: Calculations...27

Chapter 4: Efficiency...35

Chapter 5: Start by Planning...39

Chapter 6: Construction Tips...43

Chapter 7: Hooking it Up...49

Chapter 8: Following the Sun..53

Chapter 9: Lifetime and Set-Aside for Replenishment.....57

Appendix: Design Drawings, Sources, and Such..............61

About the Author...75

Other Books by Crossroads Publishing of Florida 77

Bonus Reading from HOW TO BUY A CAR........................81

Bonus Reading from PHOTOGRAPHIC COMPOSITION...93

Chapter One

General Discussion

Solar Energy. Everybody is talking about it, but few people have actually done anything with it, though some progressive power companies have started generating their own solar energy to augment what they produce from older technologies. Can you use it in your home? How does it work? Is it "here," yet, or is it still some far-off dream?

Surely you have seen those solar-operated garden path lights. That is the simplest form of a solar system. You need a solar panel, a charge controller, a storage device, and a load. The garden light contains basically a small solar cell, a battery, and some electronic components, which basically serve as a charge controller, to prevent draining the battery when there is no sun, and overcharging the battery when there is abundant sunshine; and a small LED to provide light.

A solar cell is a wafer of a particular form of silica that is photosensitive—that is, it produces electricity when exposed to light. Technically, light excites the electrons, which then move. To make the electrons all move in the same direction, two layers of silicon material are often fused together, one coated with boron, and the other coated with phosphorus. Then, since the electrons all move in the same direction, electricity is generated. There are two principal forms of solar cells, called monocrystalline, and polycrystalline. Mono cells use a silicon that is more pure, and so they are more efficient. But poly cells have been getting better, and they cost considerably less, thus you are more likely to buy polycrystalline modules. There are also thin-film, which are less efficient, and they can be bent slightly, so are frequently used for recreational vehicles and boats. On our system, we are using polycrystalline panels, though we have seen monocrystalline panels recently that are not a whole lot more money than polycrystalline panels. The voltage produced is generally uniform, usually one half a volt per cell. The current (amperes) produced is dependent on the size of the cell and the strength of the light falling on it. A solar panel is a collection of solar cells in an array. The most common size for home use is a 100-watt panel. That would normally consist of 36 cells wired in series, yielding 18 volts, which is about what you want to charge a 12-volt battery. Larger industrial panels are likely to produce two to four times that voltage and power. The bottom of one cell is wired to the top of the next cell by a flat wire called a tab wire. There are normally four columns of nine cells each. Each string of nine cells is wired to the next string by a larger flat wire known as a bus wire. You can buy all of the materials you need to make solar panels, at pretty cheap prices, until you get to the frame and glazing. That stuff is expensive to buy in small quantities, and so you will find that while you COULD build your own panels, it is a whole lot of work, and it will cost you more in the long run than buying completed panels. At the time of this writing, it is almost impossible to tell what if any, effect the tariff imposed by #45 will have on the price of solar panels, but they have been retailing for under $1 per watt. We have paid as low as $85 for a 100-watt panel, from NewPowa America Corp, and we pay for them through John's PayPal Credit account, which gives us 6 months to pay for anything over $99, with no interest. We therefore try to buy two at a time.

You then need a Charge Controller. There are two types, Pulse Width Modulation and Maximum Power Point Tracking. The MPPT is more efficient, and often considerably more expensive, but that is generally considered the better one to buy, of course, although that may not always be the case. We have found one that we consider good, and very inexpensive, listed as the CMPT02. Only the latest version at the time of this writing, is a true MPPT controller, however, so you must look for the red stripe across the center of the front cover. There is another one that is not too much more money, that also looks good to us, but we have no experience with it at this time, and that is the one in the center photo. We have ordered one each of that one and MPT-7210A to check them out, but they have not yet arrived The MPT-7210A (the one in the bottom photo) is a bit more money, but has more features, though it is only ten amps. It is, however, designed for higher voltage units, and can handle up to 600 Watts of power, so it should be capable of working with a 12-volt system, though the sellers only list it for 24 to 72 volts. We will be looking in particular to see if this one will work on a 12-volt system, but we may also decide to go to a 24-volt system for phase two, in order to keep the bus wire to a reasonable gauge. The voltage is adjustable down to 15 volts, which is normal charging voltage for a 12-volt battery. This also is the only one of the three that has a cooling fan, and does seem to be of somewhat heavier construction in general. If necessary, you could always put two 12-volt batteries in series, and hook those pairs up in parallel, (or just buy 24-volt batteries). and use a 24-volt inverter. These controllers are very reasonably priced. Others could cost up to several hundred dollars, and I am not sure you would notice the difference, as long as you don't short them out. We shorted one out like the top left picture below (the CMPT02), and were able to bridge the gap in the bus with a couple of strands of wire and some solder, and it seems to work just fine. We have had problems with the older model from the same manufacturer, however.

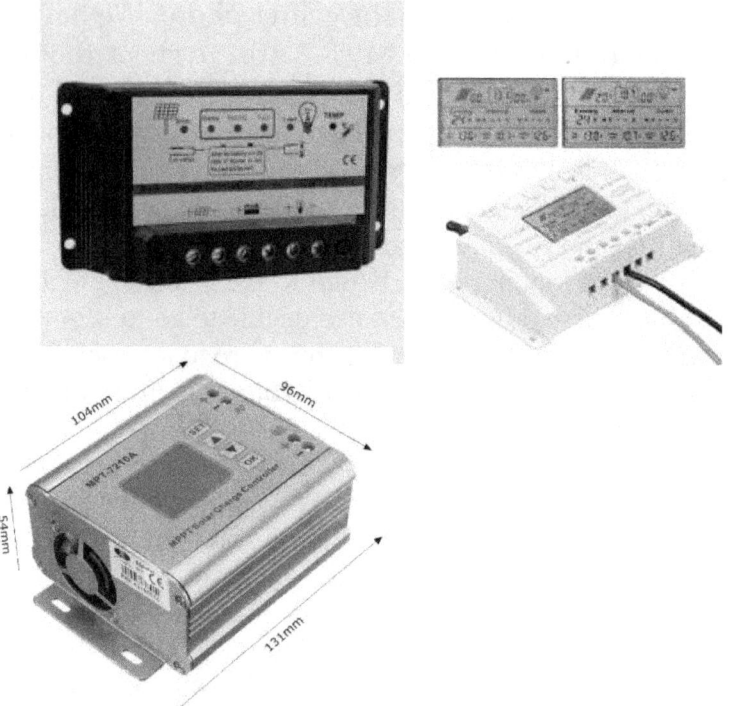

You will also need a bank of batteries. The best we have seen are AGM (Absorbent Glass Mat) batteries made specifically for solar systems, which sell for about $160 on eBay. They should hold one kilowatt-hour of electricity. You can buy a group 27 Exide battery for about $80 at Tractor Supply and other sources, or a group 29 battery from Wal-Mart that is made by Johnson Controls, for about $87 at Wal-Mart. We have all three types in our system, and all seem to be functioning well, though the AGM battery should have the longest life. The Group 27 battery claims to hold 1.260 KWH, and the group 29 about 1.440 KWH. You probably should plan on replacing batteries after five years. Solar panels lose efficiency as they get older, but still produce electricity for typically 30 or more years, losing about 10% efficiency each decade. Controllers should last a long while, but may in fact fail early, so you should keep one or two handy for replacements. So solar power is not "free,' since you should put aside enough money each month to be able to replace the batteries in five years, plus a little extra to replace the panels in 25-30 years. You can set this aside in a fairly high yielding mutual fund, as it will be untouched for some years, and the interest or yield should more than cover any inflation in the cost of the goods. $60 per month should be sufficient, unless you are using the highest-priced batteries....

We started out using 30-amp controllers, but that really is overkill, and 10-amp controllers are probably adequate, since the panels only put out 100 watts, and 10-amps at 12 volts is 120 watts, though the controller is actually handling 18 volts, which is the nominal voltage of the solar panel. We probably will use 20-amp from now on. You could use a larger one, and gang the panels together in parallel. That would reduce the number of controllers required. But the connectors required to gang the panels probably cost more than the controllers I recommend. You could use simple bus connectors, but if you buy pre-made cables with MP-4 connectors all around, that gets expensive, and it is not necessary. We also had intended to pair one battery with one panel, since at least in theory, both should be able to handle one kilowatt-hour per day. Our experience is that the batteries are not fully charged by one panel, but this location has a fair amount of shade, so we are not getting full sun all day during the winter months on our panels, and we have not had much experience during the summer months, when they should get full sun most of the day. We only had one or two panels running in the summer, and now find that we had a bad inverter, so we have replaced it with a smaller one (1000 watts instead of 1500). But we are getting a better charge with five panels and four batteries, and we will shortly have six panels with four batteries, and the new inverter, and that should work out well. As the days start getting longer, our voltage is going up. We had over 13 volts this morning (early March), which is good. Apparently, we need to cut down a lot of trees to get full sun all winter long. With the sun lower in the sky, the trees block too much sun. Phase two will be taller, and that will also help. Okay, we recently had a voltage reading of 14.9 volts, which is way too high, so we need to increase the load on the system—which we did by adding a freezer alongside the refrigerator, and using the solar system to power our cooktops, instead of running the generator for those. And it might be better to actually use one battery per solar panel, which could help use that extra energy. And it means that we really do have to cut down the tallest trees in the area, to gain more light in the winter months.

The batteries are tied together with 4-gauge stranded wire, and we use 10-gauge stranded wire to hook up the solar panels to the controllers and the controllers to the batteries. The first 10-gauge wire we bought satisfied some new spec, and was way too stiff for this use. The individual strands were heavier. It was designed to be used in high-powered car stereo systems, but is not necessary for solar system usage. And of course, we use the 4-gauge stranded wire to hook the batteries up to the inverter. When we get the larger system up and running, we will probably use heavier wire than 4-gauge, for the batteries and the inverter.

Our system is for full off-grid use, which means that we have no connection with any power utility. The power company we had to deal with was a cooperative, and they had enough connections with the legislature that there is no state control over co-ops, and they were the worst power company we have had to deal with in our adult life. They were pumping 245-250 volts through the system at all times, which is more than 10% above nominal. That means surges go at least that much above normal as well. We had to reset the clocks on the stove and microwave every day when we got home from work, and they refused to do anything about the situation. We replaced every capacitor in the heat pump, We had a brand-new refrigerator die in 3-1/2 years, and we replaced the well pump pressure switch eight times in 13 years, and the capacitor in the well pump controller once. Those kinds of problems only occur with bad electricity! But no state authority has any control over power cooperatives. We kicked them out, and bought a generator, and then started the swing to solar energy.

By the way, people have told us that it is illegal to live off-grid in Florida. That rumor comes from a deliberately misleading headline on an old Anonymous (the group) post about a lady who had been arrested for being off grid. And yes, she was in Florida. But she was in violation of CITY ordinances, not state laws. And she was charged with at least three distinct violations, and was found guilty of only one of them. She was not hooked up to the electric grid, but that was not mandated in the ordinances, and she was found not guilty. She was charged with not hooking up to city sewer as well, but she had been disconnected from that because she was not using city water, and as is true in many cities, sewage fees are based on water usage fees, so the city had disconnected her because they were not getting paid. She was found not guilty of that. She was, however, found guilty of not being hooked up to city water. Apparently the city had some idea that everyone should be drinking the same water, since it is treated with poisonous chemicals to keep everyone healthy—you know, chlorine and fluoride, stuff like that.... So they found her guilty of that. Where we are, there is no water or sewer service available, and the electric grid is not furnished by any local arm of government. Therefore everyone has their own well and septic system, and while the well requires electricity, it does not matter where the electricity comes from. There are no rules of any kind that would force us to live on the grid.

We can run the well pump and the water heater on a 5000-watt generator, with both on at the same time. We were able to run the washer and dryer as well, as long as we did not use the dryer when the well-pump might come on, or the water heater. We now have an 8000-watt inverter, and with fifteen panels and ten batteries or so, we can probably run everything we have to run, and we can always run the generator if we need more electricity. The problem then would be that we would not be able to run both the generator and the inverter on the same circuit, because they would be out of phase. If you run the generator, and the inverter is 180 degrees out of phase with the generator, you have no current at all, and you are likely to blow out either the generator or the inverter, or both. Going to a gas stove, water heater, and house heat, would drastically reduce the electricity needed overall. The average (modest-sized) all-electric home uses about 30 kilowatt-hours per day, except in periods when extreme heat or cooling are necessary.

The other possible arrangement for a solar system would be what is called a grid-tie system. This is where you feed surplus electricity into the power company grid, and they give you a credit for that. Such a system probably does not have any batteries except for emergency backup power, and it uses a totally different inverter.

One must assume that such an inverter must synchronize itself to the power curve (sine wave) supplied by the power company. When your system is not generating any electricity, you are still buying your electricity from the power company. It is cheaper to install, but of course, you still have a power bill at the end of the month, it is just smaller than what you are used to—unless, of course, you make way more power than you need, and the power company winds up paying you at the end of the month! In our circumstances, we want nothing to do with the monopoly power cooperative that runs in this area, so we are using an off-grid system. John hopes to build a much more efficient home to replace the current double-wide mobile home. But we believe that he can be fully self-sufficient with no more than 30 solar panels and 20 batteries or so. With full sun, he might be able to properly charge one battery per panel, and that would get him to the 30 KWH that the average home uses.

While interviewing John about this book, Priscilla McGee asked if you had to own your own home to install this. John opined that no, as long as you do not go into the home electrical system. That is, you could build a structure in the backyard (as long as you are permitted to do so) that could be removed when you leave, and you could just run some wires into the home, and run a few things on solar energy, so as to reduce your light bill. That was a very good question, and we thank Priscilla for asking it, and I'd like to expand on it a bit here. The largest users of energy in your home are the 220-volt major electrical items in the home, such as a heat pump, or the heating and air conditioning system, if they are separate, and, of course, the stove, water heater, and clothes dryer and a well pump if you have your own well. The fact is, it would be much easier to run solar to the 110-volt appliances. But that is not to say that you could not substitute 110-volt appliances for 220-volt appliances, particularly in the area of cooking appliances. John has a Hamilton-Beach countertop convection oven that will accommodate two 12" frozen pizzas, and even comes with a rotisserie for cooking a chicken. He also has two induction cooktops—one has a cast-iron cooking surface, which takes the energy from the induction system, and gets the iron plate hot. That means that even though this uses induction, which may be more efficient than heating a standard element, it does not require special induction-ready cookware. He has a lot of Corning pots and pans, and Pyrex baking dishes, for instance. These are useless on the larger, glass-top convection cooktop, but they are fine on the one with the cast iron cooking surface. The cast-iron one, however, consumes 1000 watts whenever it is turned on. When it reaches the desired temperature, the thermostat shuts the current off until it cools off enough to require more heat. The glass-top one, however, actually regulates the heat by using fewer watts, so it is better to use with his new smaller inverter. If you had one of those, and that counter-top convection oven, you could cook a lot of meals without powering up your big stove. You could save quite a lot on your energy bill by doing that. You could also get a window-mounted air conditioner, and leave the big heat pump off. It would not have to be a little 110-volt unit, either. It could be a heat-and-cool unit that simply mounts in the window. Maybe you wouldn't have to condition the air in the bedrooms at all throughout most of the year, and just made the living area more comfortable. That could save you a lot. If you have ten to fifteen solar panels, you could easily run a 10,000-watt inverter, and run a separate circuit into the home. You could actually cut your power bill by more than half, very easily, without going into the home electric system at all. You could run a separate wire to your clothes dryer as well (they usually plug into an outlet on the wall, and are not hard wired like the heat pump would be), and to the washer and refrigerator, which generally run on 110 volts. So there is a lot you could do when you don't own your home, which could save you a very substantial amount of your power bill. And then when you get ready to move into your own home, you can move the solar panels and batteries with you, and start saving right away in your own home as well. Good question, Priscilla, and one that John would not have thought to answer without your prompting him.

Chapter Two

What You Need and

What You Need to Know About it

So, you would need to have a place to mount panels, that is in an area of good sunshine most of the day, and most of the year. This may require cutting down a lot of trees, if they are very tall. Also, of course, the higher the panels are, the more sun they are likely to get, but then again, it gets much more difficult to work up high, and much more expensive to build higher. From a practical standpoint, building above 16 feet is very expensive and difficult, perhaps beyond the range of practicality. The panels should be mounted at an angle, so the roof of a shed seems ideal, but the roof needs to all face in one direction—which is what is called a "shed roof." Even at a 30-degree angle, which is appropriate where John lives in North Florida, which is where we are building this system, that means that the high end of the roof will be at the height of the low end plus half the length of the roof from front to back. Actually, while the one fixed number is the length of the roof, since that is determined by the configuration of the solar panels, in real terms, you would then most likely determine the length of the shed from front to back, and do a lot of calculations from that. The rise would be half the roof length, and then the base would be half the roof length times the square root of three (for our 30-degree angle). We do go into using trig functions (sine and cosine) in a later chapter, which makes this all very easy. In our case, we can build anything we want on John's property, up to 120 square feet, without a building permit. That makes the shed about 10' x 12'. Because the shed is 12' wide and the roof needs to be nearly 12' long if we build the shed 116" or 9'-8" deep, the actual rise from the low wall to the high wall, will be 67" or 5'-7", and the length of the roof from front to back will be 134", which is 11'-2". Actually, we'll allow a couple of inches of overhang on all sides, so that is about 11'-6", sufficient to hold three 40-inch tall panels with a few inches between them. The roof will be 12' wide, which will hold five 27-inch wide panels. Therefore, we can get 15 panels on one roof. To get to 30 panels—about what the average house consumes in a day, assuming we can get full sun on the panels all year round—we will need two sheds side by side. And note that the ideal roof angle, if you are not going to install a solar tracking system, is the latitude of the property on which you are building the system. That means that your roof will be steeper as you go further north (or further south, if you are south of the equator). In theory, you can get a 30% increase in power with solar tracking, but those systems cost more than adding 30% more panels, so it becomes a bit of a moot point. In fact, it might be more beneficial to track the sun in the east-west direction, if you really want to track it. That way, your panels will be receiving full sun from early morning until late evening every day. You could install a manual tracking system on the other axis, since it will vary less often than the east-west direction, and you could simply track it manually once a month or so. Even a single-axis system would cost at least about a hundred dollars per panel, unless you engineer something that will tilt multiple panels with one controller, and that would be fairly complicated—and might shade a portion of the next panel above it, for instance. Simply not worth the cost or the effort. Here is a diagram that shows the relationships of the roof angle to the location of the system. Note that if you are beyond 45 degrees of latitude from the equator, the side leg will be longer than the bottom leg. The angle required might then make it very difficult to build the roof of a shed, and you might have to resort to separate rows of solar panels.

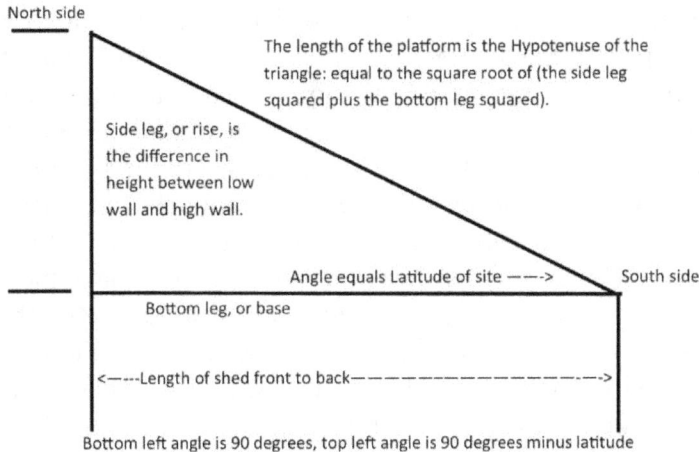

The sum of all of the angles in a triangle is 180 degrees. Since we are dealing here with a "right" triangle, the back-base angle will always be 90 degrees, so the top angle will be 90 degrees minus the front base angle—the angle that is the same as the site's latitude. If you are familiar with things like sine and cosine and tangent, you can use those to find the slope of the roof. In our case, the math is very simple, since we are almost exactly at 30 degrees north latitude. The base angle at the low (south) end is 30 degrees, so the other angle, the one at the top is 60 degrees, which is exactly double the latitude. The side leg is across from the 30-degree angle, and the base, or bottom leg, is across from the 60-degree angle. The roof, or hypotenuse is therefore exactly twice the length of the side leg, and the side leg is exactly half the length of the roof, while the base would be the rise times the square root of 3. If we were at 45 degrees north latitude, both the rise and the base would be exactly the same length. But if you can't be so lucky, you can lay it out on paper or on plywood, or something, and put in the angles with a protractor, and then carefully measure the legs and the roof, using the formula given above (in the diagram). You square the bottom leg in inches (don't mess with squaring feet and inches together), then square the side leg in inches, add those two squares together, and find the square root of that sum, and that will be the length of your roof in inches. Squares and square roots can be easily obtained using any reasonable calculator, including the ones on most cell phones.

You will need solar panels. These come in varying sizes and constructions. The lowest-price ones have no frame or substantial glazing, and are somewhat flexible, which are great on a boat or an RV, perhaps, but maybe not so good for a permanent installation, where you might want something more substantial. You can buy them in sizes ranging up to about 300 watts, but the most common one for home use is 100 watts. Most of these are for 12-volt systems. You can run them in series to increase the voltage if you need 24 or 48 volts. The actual voltage you choose will be based on the inverter you choose. When running in series, since you will have more than one, you will have to make it a series/parallel array. To do this, you gang as many as you need in series, then wire them on to the battery array that will match your voltage need, whether it be a single battery or another array of batteries hooked up in series. You would then hook those battery arrays up in parallel to feed the inverter. Since each solar cell puts out about ½ volt, and since you need higher voltage to charge a battery to 12 volts, they normally run 36 of these in series to make one solar panel, which produces 18 volts. To hook panels or batteries up in series, you must connect the positive terminal of one to the negative terminal of the next one. To hook those arrays up in parallel, you hook the negative terminals together, and hook the positive terminals together. Here is an illustration of a battery array using 12-volt batteries to drive a 24-volt inverter: If you were using 12-volt solar panels, you would probably connect two in series, and then run them through a 24-volt charge controller, and then connect them across each pair of batteries. You could also run one panel across one battery, but that seems unnecessarily complicated, and would require twice as much wire and twice as many connections, along with twice as many charge controllers, with twice as many chances for something to go wrong.

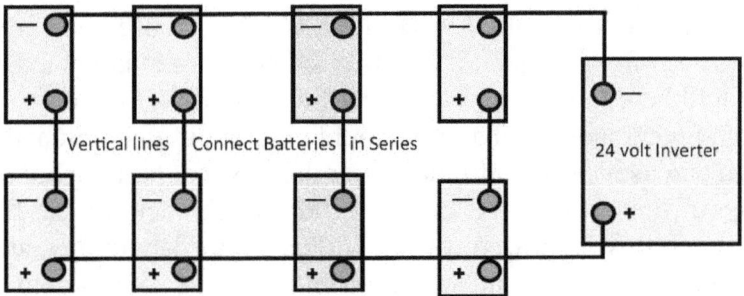

Horizontal Lines Connect Batteries to Inverter in Parallel. With 12-volt Batteries, This Array Would Send 24 volts (2 Batteries in Series) to the inverter at four Times the Rated Power (watts or KWH) (Using 4 Battery Pairs in Parallel).

Connecting batteries or panels together in series raises the voltage, but does not affect the current. Hooking them up in parallel raises the current, but does not affect the voltage. Think of Lemmings falling off a cliff. Voltage affects the height of the cliff, while current affects the number of lemmings. Or picture a water wheel spinning a generator. Voltage would represent the height of the waterfall, and current would represent the flow, or quantity of water coming over the waterfall, in gallons. A team of horses is hooked up in parallel, and affects the amount of load that they can carry (or pull). If you could hook them up in series, it would affect the speed, but of course you can't do that.

We discussed charge controllers in chapter one. We have been using one controller per panel on our system. You could gang the panels up and put two or three panels on one controller, but it seems unnecessary unless you are running a system at a higher voltage than your panels are putting out. We bought an 8000-watt inverter that converts 12 volts DC to 220 volts AC, and it seemed appropriate to run everything from 12 volts. You could, however choose to use 6-volt golf-cart batteries in a series-parallel network. That would make especially good sense if you had access to those batteries at a good price. They hold a whole lot of ampere-hours. (Amperes times Volts equal Watts, and a Kilowatt is 1000 watts. So if you had a 6-volt golf cart battery that put out 200 ampere-hours, that would be 1200 watt-hours, or 1.2 KWH.) Kilowatt-hours is the unit that your power company bills you in, and your power bill will show you how many KWH you used between the last two readings of your meter, and you can divide that number by the number of days between the readings, to discover how many KWH of electricity you use on an average day. A 100-watt panel, if it gets good sun for ten hours a day, as it should here in Florida, would produce one KWH per day. As we mentioned, a decent battery should hold about that much as well, so you can use about half that during the day and half during the night, if you want to look at it that way, but the number you get from your power bill will show you what you use in a 24-hour day, and that is the number you should strive for in designing your system. Mind you, designing an off-grid system, to get the power company off your back forever, may cause you to rethink certain things, like whether you really need to keep your sauna at that high a temperature day in and day out.... If you can cut your electricity use by 20%, then your solar electric system should cost you 20% less to build.

You will need Batteries to store the current. We had been considering using one battery per panel, but since you will use a substantial amount of electricity while it is being produced by the panels, you don't really need that many batteries. Or, to look at it differently, if we were thinking that we might want 15 KWH per day, since most of that would be consumed during the day, we would not need to hold onto 15 KWH all night long, so we would not need 15 batteries. Somewhere around half of that number would be satisfactory. When we had one battery per panel, we were seeing maximum battery voltage around 11.5 to 11.9 volts. A battery is not fully charged until it has 13 to 13.5 volts, even if it is nominally a 12-volt battery. We now have five panels with four batteries, about to head to six panels with four batteries, and right now we are seeing about 12.1 volt to maybe 12.4 volts and maybe 12.6 volts in the early afternoon, so we still need more panels to get them up to a full charge. This means that you may need fewer batteries than you might have thought. Well, now that the sun is higher in the sky, we have seen as high as 14.9 volts on the batteries, and that is way too high. It should peak at about 13.5 volts. More batteries should help, and carrying more load will also help. We hooked up a freezer alongside the refrigerator, and that has helped, and we are also using the solar energy to run some cooking appliances, rather than running the generator every time we wanted to cook something. It may be that we can find a way to create a load whenever the voltage gets over 13.5 volts. The lower-priced charge controllers have a load, but it comes on at too low a voltage, and stays on until the battery would no longer run the inverter, which is not what we need that load to do. The higher-priced charge controllers will probably let us select an on and an off voltage for the load. Having a load come on at 13.5 volts, and shut off at 13.2 volts would probably be ideal. We did pick up a fan, and that would be a reasonable load, and it helps to keep the mosquitos away, so finding a way to turn that on at 13.5 volts might be relatively easy.... When we get phase two—now looking more like phase 1-B—we will probably get an air conditioner for the shed, and that will help. But we definitely have to cut down a bunch of the closer and taller trees, in order to even out the energy levels between summer and winter. Or, we could just get some shutters to block the sun from some of the panels on the brightest days. We noticed this morning that the 1000-watt inverter would not start both the refrigerator and the freezer simultaneously. It will run both together, and start one while the other is running, but if the thermostats kick in simultaneously, the inverter dies. So we need a larger inverter, even for the smallest part of our project. The 1500-watt unit we started out with was fine, but that new 1000-watt unit is not so good. Sometimes you just can't skimp on stuff you need....

And finally, you will need the inverter and all the wiring that it takes to get from the panels to the batteries, and then to the home. And if you are running a whole-house system, you will need to ground the inverter with a grounding rod pounded into the ground. Even small inverters have a ground bus, but if you are just hooking a small inverter up to your car battery, you probably would not bother to ground it. Anything over 1000 watts, and anything that puts out 220 volts, probably should be grounded. If you are unsure what bits and pieces you need to do this, your friendly local lumber yard/hardware store should be able to advise you about that. And this book is not intended to teach you how to run electrical wires, either, and especially not at high voltages. 12-volt stuff is pretty easy to work with, as long as you have a heavy-enough gauge wire to carry the amps you need for the distance you require. There are cell phone apps that will calculate the wire gauge required for x number of amps and y distance at any specified voltage, and either AC or DC. You also need to tell it whether you are using copper or aluminum wire, as aluminum wire will require heavier wire than copper wire for the same current over the same distance. Alternating current gets a little trickier, though it requires lighter wires, and higher voltage also uses lighter wires, since it is Amperes (current) that requires the thicker wires. Any time you are working with quantities of electricity that could kill you, you should either call a competent electrician, or at least go to your local hardware store or Lowes or electrical supply store (or a good bookstore) and buy a book that will tell you how to wire the things properly. And note that a lot of generators and inverters these days that put out 220 volts, require a floating neutral. If your home is not wired with a floating neutral, you should not hook up the neutral from the generator or inverter. Use only the ground wire, and make sure the device is properly grounded! The neutral wire is the white wire on a 220-volt circuit, which is at zero volts, and which enables the 220-volt circuit to be broken into two 110-volt circuits. Floating means that that wire is not connected to ground. The power pole we were working with had the neutral connected to ground. I don't know why the generators and inverters don't do that—it may be that they know some people will not hook up the ground bus to a proper ground, and so they leave them separate. Our inverter, for instance, allows you to run only one 110-volt line from the 220 output. If you hook it up as 220 volts, you cannot run the neutral wire. We do show you one inverter in the appendix which appears to run the neutral connected to the ground, from the description. It is a lot more expensive than the one we chose, however.

Inverters are simple enough things. They take the Direct Current from the batteries and convert it to alternating current. In the USA we have standardized on 60 cycles per second, but other parts of the world use 50 cycles per second, so you must ensure that you have the proper frequency available on your inverter. Most low-price inverters use what is called a modified sine wave current. A sine wave is the shape of the curve that your electricity makes as it goes between positive and negative. A pure sine wave is a nice smooth curve, whereas a modified sine wave is actually a square wave—taller than it is wide, of course, but it has square corners. Electric motors require a pure sine wave. Computers run better on pure sine waves as well—especially if they have a conventional hard drive. Televisions prefer it as well. Anything that gets its power through a transformer is probably only seeing Direct Current by the time it gets to the device, so a modified sine wave is fine to run your laptop and charge your cell phones. But for refrigerators, water pumps, and anything else with a motor, you want to make sure you have that pure sine wave current. They are getting very tricky now about faking things on eBay, for instance. They'll call a square wave a sine wave or an M Sine Wave, to give you the impression that it is a pure sine wave, but it is not. Make sure you read the description very carefully. Anything for whole-house use should be pure sine wave. They are also advertising PWM charge controllers as MPPT, so make sure you read the description to be sure that you are getting what you need. I have had dealings with Missouri Wind, Dr. PowerJack and NewPowa America, and I am reasonably certain that those guys will tell you the truth if you ask them a direct question. Renogy is one of the oldest companies in this market, and one of the most respected, though often higher priced than we like. Some of the other guys I might not trust so much. The newest low-frequency inverters can handle a peak start power on a motor up to four times the rated power. Older inverter designs, which use MOSFET transistors, instead of a heavy transformer, are limited to double the rated power for starting a motor, so you might need a larger inverter than what you figured you needed.

    And in case you have a few days of clouds and rain, you might want to have some form of alternative generation. You do not want to mix a generator with the current from your inverter, because the sine waves can be out of synchronization. If they are not perfectly synched with each other (I assume that this is the primary difference between a grid-tie inverter and an off-grid inverter) then you will have a reduced voltage—at minimum, something like a brownout, which can damage your appliances, but it can go to zero current, and possibly hurt both your generator and your inverter. The safe way to handle this is to use the generator only as a battery charger, and run your home off the inverter. Either take a DC output from the generator, or use the generator to run your battery charger. You might want to get a small wind generator to perform this function, since it can generate current at night and in periods of bad weather—in fact, cloudy and rainy days often result in increased wind, so this would be a very good way to augment your system. You could even use a car alternator from a junk yard, to run on wind power and charge your batteries when there is no sun. We are toying with the idea of a vertical axis wind turbine to drive a car alternator for this purpose. That is something that would not be too hard to create yourself at a very low cost. You can find articles in THE MOTHER EARTH NEWS about building this kind of device inexpensively. There is a comparable magazine in England as well, and they do more of this sort of thing over there.

The advantages of a vertical axis wind turbine are many in this case. A horizontal wind collector spinning, can be very dangerous to anyone standing in the area. A sudden shift in the wind can move the blades around very quickly, so that if you were standing in front or behind it, if it spun 90 degrees to the side, you could be hit very hard by one of the blades. Vertical axis blades, on the other hand, are contained completely within the confines of the top and bottom circles, so there is no danger of being struck by the blades. They also mean that you can transport the energy from the turbine through a shaft, which could be as simple as a piece of conduit or pipe. This means that your generator can easily be located at ground level, while the turbine collects the wind much higher up in the air. This makes servicing the generator much easier. It also means that you could fasten something like an automotive flywheel to the bottom of the shaft, and spin a large number of alternators at very high speed, by putting a starter motor gear on the end of the alternator shaft so that it meshes with the flywheel. This would be a very efficient way to charge your batteries. Bringing energy down from a horizontal shaft would mean having a 90-degree converter (bevel gears) at the top of the shaft. This could also be done by using a very large belt or chain, but that would not be entirely satisfactory. Vertical axis is clearly the way to go, here.

Chapter Three

Calculations

We've talked a little bit already about calculations, but we'll try to get a little more serious here. There are all kinds of calculations involved during the construction process, but first, let's talk about how to determine the size of the system you will need. We've already mentioned that the unit of electricity actually employed by your power company, and used to charge you for electricity, is the Kilowatt Hour, normally abbreviated KWH. This unit is equivalent to 1000 watts consumed for one hour. That number can be achieved in a number of ways. You could run a ten-watt light bulb for 100 hours, or a 100-watt light bulb for 10 hours, or a 1000-watt heating element for an hour—or a 1500-watt element for 40 minutes, or any other combination of watts and time that uses that precise amount of electricity. The average home consumes 30 KWH per 24-hour day, but you can use your own electric bill to determine how many KWH you consume in a day. The meter readings at the start and end of the month will be listed on your bill. The difference between those two readings is the actual amount of electricity you used between the last two readings of your meter. Divide that amount by the number of days between the readings, and you have the average amount used per day. This amount can vary between the seasons, and can even vary by the days of the week, if you do different things on different days. You can read your own meter, if necessary, to determine the actual amount of electricity consumed on your busiest days.

Next, you should consider if there is any way that you can cut back on electricity on those days or seasons when your usage is above normal. Solar systems cost a lot less if you can even out your use from day to day, and season to season. You really do need to either rate your system for your highest-use days, or cut back on electricity use some other way—perhaps by switching to a gas clothes dryer, stove, water heater, and perhaps home heater. In our case, John has begun deriving most of his winter heat by burning wood in the fireplace. He has three acres of solid woods surrounding his home, and he can find wood lying on the ground for the most part, though he has cut down a couple of large trees to gain sunlight on his solar panels, and several smaller trees, which he is now cutting up for firewood.

Once you know how many KWH you are designing your system for on a daily basis, you have to consider what size panels you want to use. You should have some idea also, of what voltage you want to run on your Direct Current (DC) side. 24-volt and 48-volt batteries are considerably more expensive, but they store two or four times, respectively, the number of KWH as a 12-volt system, assuming that they are rated for the same amperes. Volts times amperes equals watts. Batteries are often labeled by the number of ampere-hours they can store. A group 27 battery, for instance, might store 105 ampere-hours. Since it is a 12-volt battery, that would be 1260 watt-hours, or 1.26 KWH. One consideration is that wire size is determined by the number of amperes and the length of the run. To carry more amps (or the same number of amps over a longer distance), you need heavier wire. So higher voltages reduce the ampere load and therefore can get by with a lighter-gauge wire for the same distance and the same number of watts available. DC requires a heavier wire for the same amp load and length, than AC. Aluminum requires heavier wire than copper. You can obtain apps for your smartphone that will calculate the required wire size once you supply the number of amps, whether it is AC or DC, and the distance and type of wire—stranded or single conductor and copper or aluminum. That same consideration therefore also applies to the output voltage of the inverter. Though, of course, your home is already wired, presumably, so that you probably will want to use the same voltage on the AC side as your home uses, which is normally 220 volts, at least in the United States of America. Your home may well have a 200-amp main breaker in the circuit breaker box. That means that your home is wired to accommodate 220 times 200 watts, or 44,000 watts. If your home actually used that many watts continuously, you probably could not afford your power bill. We will be using an 8,000-watt inverter, which is slightly less than 40 amps at 220 volts. Actually, it is 36-and-a-fraction (a little over 36 and a third) amps. So, we will put a 40-amp breaker in there. That is a little less than one-fifth of what the home was designed to carry as a maximum. Mind you, that 44,000 watts times 24 hours comes out to 1056 KWH per day—way more than 30 times what the average home uses. The peak power from that inverter is 32,000 watts, which covers short-duration bursts for starting things like compressors in refrigerators and heat pumps, or the well pump, or a quick surge through a heating element in the water heater. Your home does not have to have 200-amps available. You may, however, want more than the 8,000 watts that John chose to live with. Should John need more later, he can run a second completely separate system to get whatever needs that extra power taken care of. That might be cheaper than upgrading his whole home. In order to upgrade the whole home, he would have to have a larger inverter and heavier wires from that to the home, for starters. He could, in fact, feed the second inverter from the same battery bank, perhaps adding a battery or two and some additional panels. But he cannot feed the second inverter through the same breaker box, because the two might be out of phase. I once took a guided tour of a power plant with a computer club that I belonged to—the 6809 Club, if anyone remembers what a 6809 was.... (That was a Motorola CPU chip that could add two 8-bit numbers together and store the 16-bit answer by joining two 8-bit registers. It was state of the art at the time.... It was used in the Radio Shack Color Computer, and other similar devices.) The operator in the power plant threw a switch to show us just how easy it was to add power from another grid out of state. After he threw the switch, he said "Oops...." The other grid was out of phase from the grid he was on, and the voltage went down by a fraction of a volt. He assured us that he would have reams of paperwork to fill out the next day, and that he could technically be fired for allowing that power onto our grid, even briefly. He added that he did not expect to be fired for it, especially considering that most homes would never notice the brief reduction in voltage because it was so small. But it shows how serious it can be to mix AC currents from different sources. Anyway, these are some of the considerations you must consider in order to choose what you want for voltages at both the DC and the AC side of the system. And it gives you some idea how to determine the size of the system you will need. If John were to retain his all-electric home, he might not want the heat pump, clothes dryer, water heater, and well pump all running at the same time. He could easily shut the heat pump off while the dryer is running, since that only runs for a couple of hours per week. And he won't use the dryer and the stove at the same time. If he is cooking Thanksgiving or Christmas Dinner, with every burner on the stove occupied, he can shut down the heat pump as well. So, he can get away with the lower-cost system that he chose. If you want to use all of those at the same time, you should add up the power consumption of all of those units together—it should be on a tag on the units themselves—and purchase an inverter that will comfortably run all of them together, plus a few smaller things like the television and some lights. There may be some advantage to running a 220-volt inverter from a 24-volt battery array. John's system is small enough, and 12-volt batteries are cheaper and easier to hook up with everything the same, so since he was able to find an inverter that would produce 220 volts from 12 volts, he went that route. Your mileage may vary.

 We showed a little about the construction in chapter 2, but we'll go into a little more detail here. The basis for designing the platform that your panels will be fastened to is what is called a right triangle—and it is a very special right triangle, since the base is exactly horizontal, and the side is exactly vertical. The angle between the base and the platform is equal to your latitude in degrees north or south of the equator. The relationship between these three sides can be based on the functions of sine, cosine, and tangent. If you have a smart phone, you should have a calculator that can enable you to determine the size of the various components by using these. The first thing you will need to determine is how many panels you want on the platform, and then what size the platform should be to accommodate them. For this, you must know the size of the panels you will be using, which should be shown on the seller's description. You probably want a couple of inches all around each one for access. You may determine that you can go closer, or that you need more room. One factor that will determine that is how far into each panel you can get from the outer edge of the platform. Another is the method you will use to affix the panel to the platform. Will you have access from underneath? Or must you actually kneel down on the platform to remove an individual panel? In our case, phase one, the test system, has an uninsulated roof, so there was good access from underneath, and he fastened the panels using a system that was extremely inexpensive to create. Phase two will be a shed, used as an office to run this publishing business. He will want insulation there. If he wishes to utilize this same system, he will have to create insulated panels that can be fastened under each solar panel, but readily removed if it should be become necessary to remove individual panels from the roof. Phase three could be a duplicate of phase two, but it would not have to be enclosed—an open barn structure might work just as well, and that would afford easy access from underneath. If you have no practical use for the structure other than to hold solar panels, it could be just a bunch of beams crossing under the panels. Each panel could be tied to the beams in a manner readily accessible from underneath—and, in fact, you could set it up so that the panels could actually be dropped down from underneath, with no need to get on top of the platform at all. In any event, the triangle would look like this:

Here's that diagram again, with more explanations—see below

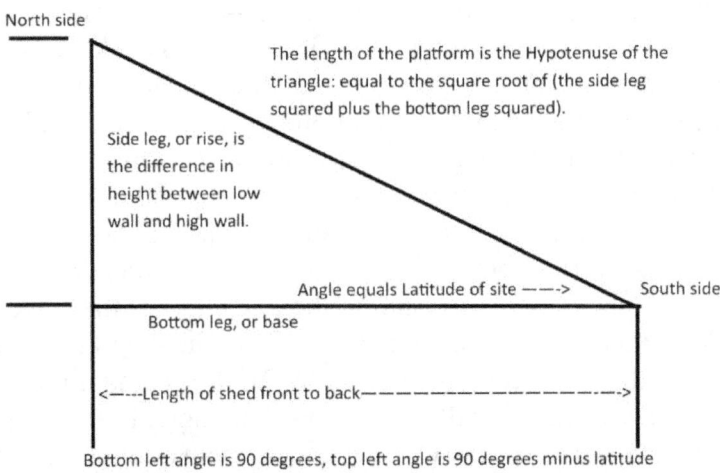

Bottom left angle is 90 degrees, top left angle is 90 degrees minus latitude

SOHCAHTOA is a helpful mnemonic for remembering the definitions of the trigonometric functions sine, cosine, and tangent i.e., sine equals opposite over hypotenuse, cosine equals adjacent over hypotenuse, and tangent equals opposite over adjacent

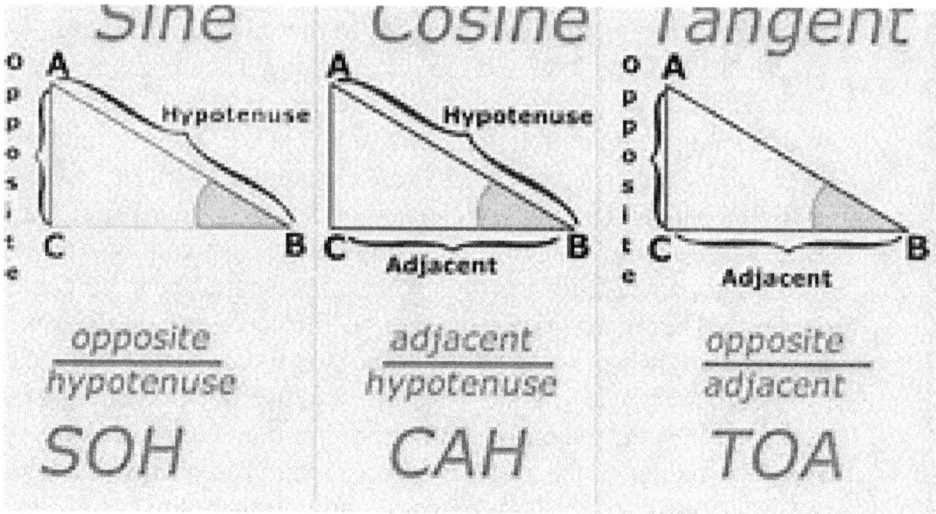

So, the Sine is the length of the vertical leg, or rise, over the length of the platform.

The Cosine is the length of the base, or bottom leg, over the length of the platform.

The Tangent is the rise over the base—and the base can be thought of as the length of the shed from front to back. Knowing the size of the platform itself, you may calculate the size of the rise and the size of the base, using the functions in your calculator. If you cut those pieces to the correct sizes, and assemble them, you should automatically have the correct angles in place, without using a protractor. Clever, huh? And you thought that you did not know anything about trigonometry! Note that these are functions of the angle itself, and we are referring to the angle that is the same as the latitude of your site, the one labelled as B in the diagram above. Make sure that your calculator is set to DEGrees, and not RADians. In our case, we are using 30 degrees as that angle. Let's say that the length of the platform is twelve feet. The rise would be 12 times the Sine of 30 degrees, and the base would be 12 times the cosine of 30 degrees. Thus, the rise is 6 feet, and the base is 10.39230 feet, or ten feet and four-point-seven inches or about ten feet and four and 11/16 inches. 23/32' or 45/64" would be even closer, but in carpentry one is rarely that accurate. In fact, in practical terms, one might actually use ¾" there (10'-4-3/4").

It would be tempting to think that the side opposite a 60-degree angle, should be exactly double the side opposite a 30-degree angle. This is not true, however, because one travels much further than other. Only in the special case of a 45-degree angle, are those two legs the same length.

There are charts available from stores like Lowes or any electrical supply stores, to calculate the wire size required to carry the requisite current over the distance needed. Luckily, you don't need to buy the charts and learn how to use them, because there are apps available for your smart phone to do that job. You tell your phone whether it's AC or DC, and how many amps you need to carry, and the distance, or length of the wire and whether it is single-conductor or stranded and whether it is aluminum or copper, and the app tells you exactly what gauge wire to buy. Most electrons move along the outer edge of a strand of wire, so solid wire must be considerably heavier-gauge than stranded, to carry the same load over the same distance. Similarly, a wire made from heavier strands, will need to be a heavier gauge than a wire made from thinner strands of individual wire. You will note that our 4-gauge wire was made from very thin strands of individual wires, so it may have been more efficient than a wire made from heavier strands. You may have to choose to use separate apps for DC and AC wire sizes. There are apps made for marine use that specifically reference DC applications.

Chapter Four

Efficiency

Unfortunately, the efficiency of such a system will not be 100%. The panels themselves lose some energy, the charge controller will lose some, the batteries lose some in the charging process and a little more while in storage, and the inverter will never be 100% efficient.

The biggest single area of concern, however, is the aging of the panels and batteries. The panels will lose efficiency at the rate of approximately ten per-cent per decade. And the batteries only have an estimated lifespan of about five years. Most panels have a 25-year warranty, with free replacement up to perhaps ten years, and pro-rated after that. Most batteries will have one-year free replacement, with a pro-rated plan for two or three years. Some batteries may have up to three years free replacement and pro-rated up to five years. Such batteries would usually cost considerably more, however.

But you must also consider the loss of efficiency in our current system. Generators are not 100% efficient, either. And transmission of power over miles of cable is not efficient. The step-down transformers bringing the current from 440 volts, at which it is transmitted, to 220 volts coming into each home are another loss of efficiency. And all of that equipment has a lifespan as well. Those lines and transformers and poles need to be maintained, and occasionally replaced. When they move wires underground, working on them costs a whole lot more. The fact is our current system is woefully inefficient, and we all suffer every time that system goes down. And on occasion, people are left with no electricity for a very long time. If you own your own power system, you control what happens when the system has a problem. If you installed the system yourself, you have some familiarity with it, and you may be able to do all of your own service. Certainly, you should have some sort of gas-powered (or diesel or propane) generator to replace the whole solar system at least for your most necessary uses, like your well-pump and/or water heater. You should have a few spare charge controllers lying around. If you can, you should have a spare inverter, so you can change it out on short notice. You should at least have a smaller inverter that you can toss in there for short term for emergency use, that will run most of your needs. Or perhaps you might choose to have a Modified Sine Wave inverter as a backup, since they cost considerably less, and you might only need it in there for a few days until you can get around to picking up a new higher-quality inverter. You might then have an uninterruptible power supply that you can plug in to run your computers, and other sensitive devices. (A UPS is essentially a small inverter with a built-in battery as a power supply. So, you can plug the UPS into the Modified Sine Wave inverter and use the UPS to power your sensitive devices.)

You might also consider having two separate lines with two inverters, so that some of your home is usable for an extended period with no actual work being done on the system. John is considering a smaller home than his current double-wide mobile home, which is 24' x 40'. He wants to go to a home that is 16' or 20' by about 32', with two stories. The first story would have the master suite on one end, and a great room, with the kitchen and dining area against the opposite end of the home from the master suite. Upstairs would be a media room, with a full bar and mini-kitchen, with about a 5' x 9' movie screen at one end, and an office at the other end. The office could be used as a second bedroom, and would include a ¾ or full bathroom, accessible from the office or the media room. The master suite would have a full bath, but next to that would be a half-bath for guests that would be accessible from an outside deck or from the inside of the great room. There would be four separate temperature-controlled zones: the great room would be one, the master bedroom another, and the office would be another, and the other one would be the media room. Typically, only the room he was using would be heated or cooled. They would use 220-volt wall-mounted heat/cool units with remote controls. None of them would be anywhere near as large as his current heat pump. That would dramatically cut back on electricity use. Since the upstairs would be usable as a completely separate living unit, such an arrangement would permit the two floors to be serviced by different electric systems altogether, with two inverters. A generator could be hooked up to the well pump for emergency use of that, or he could throw a switch to move the well pump from one system to the other. More likely, he could just turn one breaker off and then turn on the breaker in the other system. The washer and dryer would be on the downstairs system, but he could use a public laundromat in an emergency, if that one ran out. Or he could run separate outlets from the upstairs system to the washer/dryer bay. Both inverters could run off the same battery bank. Any individual panel, charge controller, or battery could be removed from the system until it could be replaced. This kind of system would almost never be fully disabled. When you design your home to work with a solar system from the git-go, you can work safeguards like this into it.

Any perceived individual inefficiencies can be worked around by building additional capacity into the system during the design stages. For instance, if you planned on replacing the solar panels after 25 years, you could start with 20% more than you thought you might need, so that even with reduced efficiency, you would be covered with all the electricity you originally calculated that you would need, right up to within five years of the time you were planning on replacing the panels.

Batteries should be replaced every five years, or you could let individual batteries run until they die, but that would result in more emergency repairs. You could get some used batteries to start with, and then set them up for replacement at 20% of the battery pack each year. We have, for instance, seen deep-cycle batteries at our local county auction of computer equipment. These used batteries were not yet in the end-of-life stage, but they had been replaced before that in order to reduce unusual maintenance. These were external batteries that were part of an uninterruptible power supply system. That would reduce your annual outlay. You should set aside some money every year, to replace these. If you had fifteen batteries, which would store about half what the average home uses per day, so that would be a minimal system if you are retrofitting, you should set aside at least the cost of three batteries per year—perhaps as low as $20 per month. This money could be set up in a high-yield mutual fund, or a CD with your local bank or credit union. At the end of five years you would have enough money to buy all the batteries together. Hopefully the interest would cover any rise in costs of the batteries, but on the other hand, the cost of batteries has been coming down. In fact, if you really wanted to go green, a lawn could be used as batteries. Just google "earth batteries" if you want to get into something like that.... Be aware, they take a whole lot of earth! If you set aside twice as much money as you would need, you might get to the point where you could cover the cost of one year's batteries with just one year's worth of gain in the mutual fund or CD. You would then have a perpetual fund set up and would not have to contribute to it again. Much the same can be done for the funds to replace the panels, where you have a lot longer to prepare for the expense.

Chapter Five

Start by Planning

The first thing you need to do is to decide what sort of edifice will hold your panels and the electronics. The panels could go on the roof of your home if you are building from scratch, and can match the roof angle to the latitude of the home. If you are doing this, you might want to make some metal brackets that would hold the panels a couple of inches above the roof. This would have to be strong enough to withstand the most severe weather you are likely to encounter. But if you run attachments through the roof, you should be able to guarantee that weather cannot come into your home through any holes in the roof.

Next you must decide how many panels and batteries you are going to need to cover your electrical needs. Then you need to lay out a document showing how the panels will fit on the roof, and hence, what size structure you envision to hold the panels. The more panels you have, the more you must be certain that you can easily service them. Can you get to them from underneath? You will notice that most professional systems arrange the panels on steel racks. For a large system this is by far the easiest system to maintain. You can access each panel from underneath. Our system will be on the roof of a shed, and if needed we may duplicate that with an open shed instead of a fully-enclosed one, but more like a small carport.. This will require accessing the panels from the top. They will be mounted directly to the roof of the shed, which means that we will be caulking the underside of anything that is attached through the roof. We want to keep the weather out of those screw holes. This is the easiest system to build, but the hardest to access for repairs.

Finally, you are ready to design the structure itself. What type of construction will you use? You can use steel studs with wood sheathing, all wood, all metal, or just a rack like the professional installations use. Such a rack could be made of metal, but could be done for probably a lot less money, on a small scale, with wood. Some of this may depend on your familiarity with working with various materials, and the tools you have available.

Then you can lay out your plans on paper, after all of the details have been finalized. One consideration will have to be how much room you need between the panels, horizontally and vertically. Will you have to kneel on the roof between the panels to hook them up? We will be using a 3 x 5 grid, with the panels arranged vertically (not that they will be vertical—just that the longer sides will be aimed up and down the roof, while the shorter sides will be laying across the roof.) The array will be five panels wide by three high. They will be fastened by blocks of wood that will be screwed to the roof from underneath. We will put them in place, then screw up from under the roof. If we decide we can drill holes precisely enough, we could use carriage bolts going down through the roof and then nutted to the roof from underneath. That would be the most secure, but a lot of effort to line them up. We could create a template from paper, or cardboard, or preferably hardboard (the generic name for Masonite.) That would be a workable solution. We would put the panels in place from top to bottom, one row at a time, starting each row in the center, and then moving to the sides. In this fashion, we would not have to kneel between the panels, because we can reach the top row from the back of the roof, and carefully reach across to get to the center row for service. Service won't be fun, but it won't have to be done often, so it should work. We would probably get one of those lightweight scaffold units from Harbor Freight or Tractor Supply to facilitate that. Initial installation, and ultimate replacement after 25 years, which would involve complete removal and reinstallation, with removal starting from the bottom, so we can walk all over the roof while replacing the whole shebang.

Just be sure that if you design this yourself, that you have all of the components sized to meet the needs of the structure in terms of load, severe weather and so on. Again, our system is small enough that 2" x 4" studs, with 2" x 6" joists, and 7/16" OSB or CDX skins, should be adequate. (9/16" 5-ply plywood would be a preferable surface material.) If you are working alone, you might find it easier to work with Tongue-and-Groove Decking. These are boards that are usually 5/4" thick by 6" wide, or possibly 8" wide, in standard lengths up to 16' long. These would be easier to handle than 4' x 8' sheets of plywood or OSB. They hook into each other for strength, and would certainly be adequate for the loads expected in this structure. These are just standard construction materials, available at any lumber yard. Larger units, with greater unsupported lengths, might require heavier construction. You might have to consult an engineering firm or architect to get it right. We were going to use very short walls—5'-4" on the low end, and ten feet on the high end, but experience with the first phase changed our minds, and we will be going with 8' walls on the small end, and 13'-8" walls on the high end, so the panels get more sun in the winter. The higher building may require slightly heavier construction in the back, with perhaps some bracketing or trusses or gussets thrown in. We might even go higher, by building the original planned heights as a second story on top of an 8' high first story.

Then you have to devise a shelf or loft to hold the electronics. We ran the wires through the roof for phase one, but we are considering running them over the top of the roof for the main building. This will improve weather-proofing, but will require that we clip the wires to the panel frames as we go up the roof, to keep them from whipping around in heavy winds. If you can keep the wires short enough, 10-gauge stranded wire should be sufficient to hook up the panels to the charge controllers, and the charge controllers to the batteries. We are considering using heavy copper pipe or steel conduit for the low-voltage buses, but may go with heavy wire instead, using 4-gauge or heavier stranded wires from the battery to the bus. Using open buses creates the danger that something could accidentally create a massive high-ampere short circuit, but really, nothing should be able to get close to the buses—if they are separated enough—to short them out. And having open buses makes the work of connecting the batteries to the buses, ever so much easier! Or we could do it like phase one, and just make up short individual wires to go between each battery, and hook two up to each connector. Or we could use heavy-gauge insulated wires, and put each connection into a standard electrical box. The bus would then be cut and bared inside each box, the wire from the battery would come into the box from the bottom (the bus would come in from the sides), and they would be clamped together in some fashion—you probably can't buy those screw-on plastic junction things in a large enough size for this part of the project.

You definitely want the inverter right by the batteries, so that you are transporting Alternating Current from the solar system to the home itself, unless the panels are right on the house roof. Direct current over any long distance would require extremely heavy wiring. It is much easier to run long distances through AC than through DC! And even if you are just setting up a small camp, and will not have any 220-volt appliances, it is easier to run the power over a 220-volt line, than over a 110-volt line. Using a breaker box makes it very simple to break the 220-volt line into two 110-volt lines. John bought a 30-amp breaker box from Lowes for under $30 to power up his motorcycle repair shed.

When you have resolved all of these quandaries, and you know exactly how you are going to do each part of the job, it is time to draw everything out on paper, so that you can create a materials list, not only for the construction materials, but also for the electrical supplies you will need.

Very little of the plans would need to be full-size, unless you have a great deal of difficulty envisioning how things will fit together. John drew the plans for phases one and two on his computer, using Microsoft Publisher. You can do simple sketches with pencil on paper, and then use your cell phone to convert to a pdf document if you'd like. CamScanner is the app for that, and it is available for IOS, Android and Windows systems. It is available from Int Sig, and available from the App Store, Play Store, and Windows Store, in both free and "pro" versions. It lets you use your phone's camera as a scanner for documents. Every time I use it, I think of those German WWII spies, with their Minox cameras, taking pictures of US/Allied documents, and needing eight days to get grainy black and white prints back from the lab. I can have full-color high-definition, high quality .pdf documents available within seconds, ready to send anywhere on the globe that has an internet signal. You can also use any of those home-design programs that are available for your laptop and desktop, and actually for an Android tablet as well, and probably IOS also.

Some of John's plans are included in the Appendix to this book.

Chapter Six

Construction Tips

You will need a certain minimum collection of tools. This would include at minimum, a circular saw and a reciprocating saw, a drill, and preferably an impact driver for driving long screws. You can generally find a four- or five-piece battery-operated tool set with a drill, an impact driver, and a circular saw and a reciprocating saw, with a couple of batteries and a charger, in the $100-$250 range. John built phase one of this system with one of those kits that he got on sale. The circular saw was only a 5-1/2" trim saw, but that will actually cut 2 x 4" lumber, as long as you stay at 90 degrees. (You can miter it, but not bevel it.) You will need measuring devices, which should include at minimum, a 25' tape measure, a level, a protractor or other angle finder (the electronic levels with an angle feature are great) (we also list some clinometer apps for your smart phone in the appendix) and fasteners, which could include a hammer—at least a 20-ounce claw hammer should be available at all times—but an air-operated nail gun with a decent compressor would be great. It would not hurt a bit to have three nail guns, one for studs, one for roofing, and one for finishing nails and tacks. John put phase one together with all deck screws, using Torx-head fasteners, (sometimes called star-head) because they really do work better than Phillips-head when the going gets tough. The drill and impact driver came in handy for those. There can be no doubt that screws are better than nails, but for the shed we'll be using for phase two, nails would certainly be adequate. John likes to use Liquid Nails and screws for this kind of stuff. Liquid Nails is a construction adhesive that comes in a tube that dispenses from a caulking gun.

You should probably build a sturdy deck to support your structure, though for certain structures, anchoring with 4" x 4" pressure-treated posts inserted into the ground would be easier, and that is what John did for phase one. The deck structure could be based on concrete piers cast into the ground, or simply on 16" x 16" x 4" concrete blocks laid on the ground and leveled with sand. If this is done, the structure ought to be tied down with screw-type tiedowns. Normally, you would just dig a hole with post-hole diggers or a 4"-6" spade, put the screw augers in the hole, and then backfill. Then you would use lag bolts to tie into the base of the structure. There is a photo of this type of tie-down in the appendix.

Whatever you base it on, it must be level and plumb throughout. John has levels ranging from torpedo levels, which are 9" long, up to a 6' aluminum level. You can always use a length of lumber that is absolutely straight, if you need to run a level over a longer distance. You can sight down the top of a 2 x 4 or a 2 x 6 to be certain it is straight and true. It can be warped a little on the short edge, because you want to turn it turn it so that it is taller than it is wide, and sits on the narrow edge. Then you just put a level on top of that to see if the two things the wood is resting on are level. You can also use a water level. Take a length of transparent tubing, set one end at the end of the comparison height, run the other end over to the other thing you are comparing to the first end, and then fill it with colored water (food colors work well) until the water in the first end is at the level you wish to compare it to. Since water will come to rest with both sides at the same height, you can check the height of the second end very accurately this way, no matter how long the distance in between them is—within the practical limits of how much hose you can obtain, of course, but hundreds of feet, in any event, more than you should ever need for projects like we are discussing here.... And the color is simply to make it easier to distinguish. If you can see the top of the water in the hose, you're good to go, color or no color. And the hose should always be below the level you are trying to match, just to be on the safe side. And you don't want any air bubbles in the hose, either. If you are using blocks, set the first one, make sure it is level in both directions, and then set the others, and level them to the first one, and make sure that they are level in both directions, and then recheck the level between them and the base block.

Then you can lay 4" x 4" pressure-treated lumber across them— or 4" x 6" if you need the extra strength—that should be the same length as the structure. You will need two or three block-pairs and beams, depending on the width of your structure. Then you lay floor joists across the tops of those beams, 16" apart and at 90 degrees to the beams., probably made of 2" x 6" Pressure-Treated lumber, though a smaller structure could get away with 2" x 4" joists. You may also determine that you do not need pressure-treated lumber for the joists, since they should never be in contact with the ground. Those joists should be 3" shorter than the width of the structure. You then tie the joists all together with another piece of the same lumber, the same length as the complete structure will be, running across both ends of the joists. The joists should be anchored to the beams, either by toe-nailing (or screwing), or by hurricane clips. The end runners should be either screwed or nailed into the ends of the joists. That finished structure should be exactly the size of the floor of your structure. Then you just cover that with plywood or OSB. Floors should be using 9/16" plywood or ¾" OSB to ensure strength, but if you will not be putting any heavy loads on the unsupported areas of the floor, you could get away with 7/16" OSB. Joists should be 16" apart, unless an engineer has certified that you may use 24" between the joists.

Once the floor is in place, you build the stud walls. These are normally built intact on the floor, and then elevated into position, and held vertical with pieces screwed from the middle of the stud to the middle of the end joist (you would normally build the longer walls first.) In other words, you put a scrap piece of stud at a 45-degree angle from the center of the outside joist to the center of the outside stud on both ends of the wall. That holds the wall in place at a vertical angle. Then when you elevate the shorter walls, you screw or nail the shorter walls to the longer walls. Once all four walls are in place, you remove the scraps that held the two walls vertical. Normally, the longer walls have the end studs doubled up to provide a stronger joint for the shorter walls to fasten into. Note that we will be using a shed roof, so there will be a height difference, depending on the angle the panels need to be at. In our case, the wall on the north side of the structure will be about six feet taller than the wall on the south side. There is one footer on each wall, and a double header. The second header on the longer walls will be shorter than the first one. The narrow walls will be built on the floor with only one header at the top. Once the walls are in place, you would add the second header on top of the shorter walls, which would be long enough to cover the ends of the longer walls that were left bare by the shorter top headers. All stud walls should be nailed into the joists and the end runners through the plywood floor.

Now you have to make the roof rafters. Either these will be fly-cut to tie into the walls if the walls were made horizontal at the top, or you can make the tops of the north and south walls at the angle of the roof, and just fasten the roof rafters on to them with hurricane fasteners. That is how we plan to do ours. To do this you should cut the top of the rafters on an angle and then bevel the headers at the same angle so that the headers form a straight extension of the vertical lines of the studs. The side walls will be made with the stud ends beveled so that the header fits squarely across each stud as it progresses from the tall wall down to the short wall.

Then the roof will go on top of the rafters. Rafters are normally standardized at 16" apart as well, which gives adequate strength to the roof, which should be made of the same material as the floor, preferably 9/16" plywood.

Because John will be using the shed as his office, and it will have all the comforts of home, he will be insulating the roof. The best way for us to do that is to create insulated panels that will fasten to the roof joists, about the same size as the solar panels. That way, we can remove any solar panel by simply removing the insulated panel inside the room, and remove the screws holding that solar panel into place. Then we can disconnect the wires from the panel at the MC-4 connectors, and then lift the panel off of the roof.

We fastened the panels to the roof of phase one in a very unique fashion. They sell Z-shaped clips to hold the panel to the roof, but those clips add more than 15% to the price of the system. We took pieces of 1" x 4" lumber, about a foot long, placed them in the corners of the solar panel frame, and screwed them to the frame using the screw hole provided in the frame, or in some cases, drilled a new screw hole, when we cut the pieces shorter. Then, when we had the frame in place, we just figured out where the corners were, and ran screws up from under the roof, into the blocks of wood. With the wood screwed securely to the roof, the panel was held securely in place at each corner. The single screw we placed through the frame really doesn't do anything except hold the wood blocks in place while we place the panel onto the roof. After that, the screws up from underneath—we used three in each block—provide enough force to hold the panel down through any extreme of weather, and they maintain the panel in position both horizontally and vertically, since the wood blocks are touching both the top and bottom frame, and the side frames, at each corner. We got at least 8 blocks out of each 8' length of 1" x 4" lumber, which costs a couple of bucks, and at four blocks per panel, we got two or three panels out of each piece of lumber (eight inches of wood is plenty enough to hold the panel in place at each corner for a yield of 12 blocks per stick of lumber), so our cost was under a dollar per panel. With the 8" long blocks, you have to drill a hole closer to the end in each side of the frame, to secure the block to the panel while installing the panel. If you want to use this method to hold the panels onto the roof, make sure that the electrical boxes on your panels do not extend beyond the back edges of the frames. The panels we buy from New Powa America are panels which are specifically designed with a higher frame, so that they can be mounted flat. If you want those when you place the order, you must specify that you want the taller frame, otherwise you will get the shorter frame, and might have to shim the aluminum frame up off the roof by about a quarter of an inch. Because this method for mounting the panels creates some additional concerns about access with an insulated roof, we may choose to spend the money for the industry-standard mounts, for phase two. If we make an open shed for phase three, we are likely to fall back on our lower-cost system for that unit.

Chapter Seven

Hooking it Up

Okay, now you've built the shed, or other structure; how do you wire everything together and get it to work? You'll need lots of wire—how much will depend on just how large your system is, and how far the panels are from the charge controllers, and how far the batteries are from the charge controllers, and how far the inverter is from the batteries. All of those distances, but especially the last one, should be kept as short as possible. The last one carries the most amperes, so that is why we want to keep it short. The more current it carries, the heavier the wire has to be, and the further it travels, the heavier the wire has to be. If you can keep it to 4 gauge stranded wire, the wire will be a lot more flexible, and a lot easier to work with.

ALL CONNECTIONS SHOULD BE SOLDERED! Because we are working outside, we found we needed John's 400-watt Blue Point soldering gun, straight off of the Snap-On truck, and not cheap. You might find one of these used on eBay for not much money, or you may be able to find something else suitable at someplace like Harbor Freight or Tractor Supply, or a decent local electrical supplier. But 75 watts probably will not cut it. If you don't know how to solder, proper technique is important. You don't melt the solder with the gun. You heat the work with the gun, and then touch the solder to the work. For the solder to flow properly, the work must be hot enough to melt the solder and keep it melted until it fills the voids. Failing that, you'll wind up with what is known as a "Cold Joint." It will not be physically as strong, and it will not conduct electricity as well, either. See if you can find a video on YouTube or Vimeo, that will show you the proper way to solder, if you are not certain you know how.

And okay, of course, you are not going to solder the connections with the charge controller—though you COULD "tin" the wires with a thin coat of solder, for better results, though we found that that was not only unnecessary; with our controllers, the available room was not sufficient for tinned wires. 10-gauge wires are the minimum I would recommend, and the connectors are designed to barely accommodate that size wire. The solar panels themselves will come with what are called MC4 connectors. These are waterproof connectors. You will have to either purchase wires with an MC4 connector on one end, or build your own. We did, in fact, build our own. There are videos on YouTube that will show you how to connect those connectors to the wire. You need a wire with the MC4 Connector on one end, and bare wire on the other end, to go into the connectors on the charge controllers. We do NOT recommend that you use the extra-heavy 10-gauge wire that they sell for high-powered audio equipment. It will not fit as well, or be held as securely, in the charge controller's screw panel. Standard 10-gauge stranded wire is far better for this purpose. Measure all distances carefully, and make the wire just a couple of inches longer to allow it to curve nicely, but do not make it so long that it hangs down and has to come back up to get where it is going. And if you determine that you need to use 8-gauge wire from the panel to the controllers, you will not be able to fit that into the controller's screw panel. The best solution is to find some sort of pin that you can solder on to the end of the wire, that will fit into the controller.

Then you'll need wires from the controllers to the batteries. Again, the end going in to the controller will need to be bare, or with a pin soldered on the end. Most of the batteries have a screw-type connector on top of the battery, though they may also have a more standard automotive connector. You can use a standard battery cable end connector that has a screw clamp on the top (if your battery has posts) or use the screw fixture provided. You should solder an end fixture onto the wire that will fit properly over the screw. Especially when you have multiple wires connecting on to the battery, you should not just wrap wires around the screw.

These are the kind of fittings you need to solder onto any wires that will hook up to the batteries using any kind of screw connectors. Match up the wire gauge and the screw hole size to the wire and screws you will be using. You can buy these from eBay or any automotive store or electrical store. You can also find them at any home center like Lowes.

Then you need to connect the bus to the battery. The simplest method may just be to use short lengths of heavy wire from each battery to the next in line. Thus, you would have 3 connectors on each battery terminal. Another solution would be to use a heavier bus— especially if you have a whole lot of batteries. Then you can run the 4-gauge stranded wire from the battery to the bus. You would then have only two connections on each battery, the controller and the bus. Of course, you need a bus for the positive side, and one for the negative side as well. You can mount a standard plastic electrical box on the wall, run the heavy wire through it, cut the wire, strip the ends, run the battery wire in to the box, and use a twist connector to tie the whole thing together or some other means to clamp them together if you can't find twist connectors large enough. If you choose to use a solid bus, you can just bolt the battery wire to it, using another connector like the one that fastened the wire to the battery.

If the bus is simply wire, you can connect that directly to the inverter. If you used a solid bus—not the best idea, even if you think nothing can short it out, that is a whole lot of current, and could cause one hell of a spark, and start stuff on fire—but if you did, it's a simple matter to connect the bus to the inverter with a short length of cable, making certain that it is substantial enough to carry the current that the inverter can handle. For instance, if we are using 12-volt batteries to supply an 8000-watt inverter, we need to be able to handle 667 amps. 3/0 wire could handle about half that capacity over ten feet, for ten minutes. You would need multiple strands to accommodate the full capacity. That would take one hell of a cable, even going just 10 feet! And your bus should be about as heavy as the wire from the bus to the inverter. This is why we want to keep distances short—especially between the batteries and the inverter. It is also a very good argument for running higher voltages on both the DC side as well as the AC side! And you might consider the likelihood that you will ever run the inverter at full load, but unless you shut some things down, there is a very good chance that at some time, it will draw that full load, especially if your well pump and water heater come on at the same time as your heat pump, for instance....

There is an app called LEARN ELECTRICAL WIRING available for Android phones, and probably something similar for IOS. There are also specific marine apps for calculating wire sizes for DC systems, One such calculator we used (Blue Sea) said we could get away with 6-gauge wire for our phase 1 system, where we used 4-gauge wire, but there is no harm in playing it safe. Actually, we initially had a 1500-watt inverter, which was not working well, so we got a new 1000-watt inverter, so it would probably call for 4-gauge for the 1500-watt inverter....

Chapter Eight

Following the Sun

We mentioned this in an earlier chapter, but we'll say a bit more about it here. You can achieve about a 30% increase in power if you install a solar tracking system, and let the panels track the sun from morning to evening, and through the various seasons. That would require a dual-axis tracking system, that would both raise the back end up in the winter, and swing the panel through an arc from morning to evening. That could be pretty complicated. Getting a dual-axis system for each panel would cost more than the cost of another panel. And if you buy 30% more panels, you get the same effect, at less than 30% of the cost. My feeling is that it isn't worth installing such a system. Tracking the sun from morning to evening is much more important than tracking it through the seasons. Even a single-axis system, however, is well over half the cost of another panel. The only way to do this and make it economically feasible would be to use an automatic tracker to move a whole row of panels. It would be extremely difficult to control more than one row with such a system, though it is probably not impossible. It would make access to any given panel unnecessarily difficult, should service be necessary later, however. Another problem is that any dual-axis system would require that all four sides of the solar panel be able to move up and down, and that means that your physical connection between the batteries and the structure is necessarily much weaker than a solid connection would be. With a single axis system adjusting only for seasonal differences, the bottom could be held solidly with two hinges, and that would be nearly as strong as a fixed mount. But the sides would both have to be capable of moving up and down for the panel to follow the sun from morning to night, and that would necessitate a much weaker connection to the structure—not good for high winds or fallen branches. Again, just one more reason not to track the sun with your system.

I suggested earlier that it would not be difficult to devise a manual tracking system that would raise and lower the back of the panels on a seasonal basis. You could create a beam (wood or metal) that would fasten to the back side of each panel, and mount the front side to the structure with hinges. You would then have to put a threaded rod (an acme thread, like a screw-type car jack would be best, because they support more weight.) at each end of the beam. The bottom of the threaded rod would have to be able to swing fore and aft (not left to right, but north to south) by at least a few degrees, because as you raise the back of the panel, the screw would have to swing to a forward angle. You could do it with a single screw in the center, if the screw were capable of withstanding enough weight, but that would be difficult to get to for a manual system. If there are two screws on the ends, you could get to them easily. One in the center would work if you had a motor mounted there that could turn the screw via an electrical switch. Such systems do exist, but again, it would only add to the cost.

If you really wanted to have a system that tracked the sun from morning to evening, it would not be impossible to control each row from a single one-axis solar tracker. Each panel would have to have a cam under it, which could be made from plywood. The north and south ends would both have to pivot from the center, at a point several inches above the structure. Then as you pulled on the cam from underneath, it would cause the panel to tilt into the sun at all hours. That would have to be automatic, since it requires constant adjustment. The seasonal adjustments could be made once a week, or even once a month, and you would never be far off—and if you missed December and June, you still would only be 5% off. That's not enough to cause much sweat. But the pivoting arrangement for the daily tracking system would greatly reduce the strength of the system—the ability of the system to withstand extremes of wind and other weather-related phenomena. There might even be a far greater chance or extent of damage, should a branch fall onto the solar array.

All in all, for a home system, it would seem like a much better idea to just work in a factor of an additional 30% capacity to cover the loss of efficiency. That would give you a little more capacity in the summer, when you might expect to need more electricity for cooling. And don't forget to add another 25-30% for deterioration over 25 to 30 years. You'll still spend a lot less money, with a lot fewer headaches, and have plenty of capacity for as long as you'll need it.

Chapter Nine

Lifetime

And Set-Aside for Replenishment

We already went into this a little bit, but Solar Energy cannot be considered completely free after you have paid the up-front costs. The devices you need—solar panels, charge controllers, batteries and inverters—are not free, and do not last forever. That said, the panels should outlast the length of time that most people own their home, charge controllers are inexpensive enough that you should keep a few extras around, you only need one or two inverters, so you might keep at least one of those on hand. That leaves batteries as the major cost of replacement. If you have 15 batteries, not quite enough to keep the average all-electric home lit up all day and night, but possibly enough if you switch to gas appliances where available, you could have them on a schedule where you replace one fifth of them each year. That gives you four months to save up the cost of a single battery, roughly $80 to $300 or so, each, depending on voltage and current capacity. That works out to about $20 to $75 or so per month for batteries, and maybe another $10 to $20 or so for replacing solar panels every 25 years or so, at $1500 to $4500 or so over 300 months, and leaving enough for the occasional charge controller and inverter. If you can afford to set aside double those figures, you might be able to build up a fund over a decade or so, where the interest on the fund itself would make those payments for you. Naturally, you would want to keep these funds in a relatively stable and well-managed mutual fund that yields higher returns than your local bank.

Proper ventilation is a must for the charge controllers, batteries and inverters. This will increase their lifetime considerably. Inverters, especially, get hot. Charge controllers get hot when the panels are really cranking out electricity, but if you get controllers that are rated higher than what you need, you shouldn't have to worry about them.

Do make sure that they are well protected from the weather. We think this one got fried because water got into the screw terminals. It was hanging upside down just inside the door of the battery compartment. They should always be fastened securely to a panel (which could be a 2 x 4 fastened at the top of the battery compartment) and the screw terminals should be on the bottom, so that water cannot get in to that area.

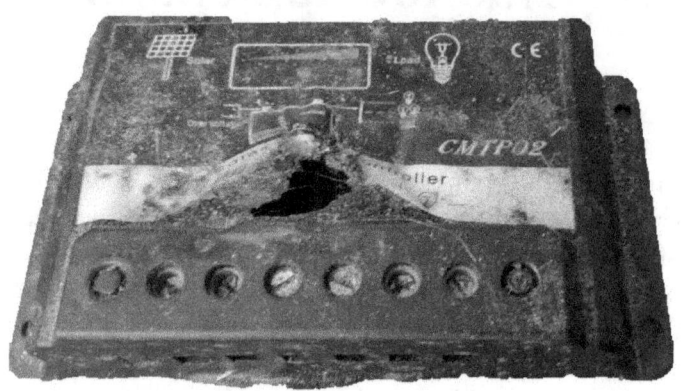

They are fairly reliable otherwise. In fact, John shorted one out completely, and was able to repair the break in the internal bus, and use the controller normally. The solar panels themselves are fairly sturdy, but can be damaged in high winds by being hit by rogue tree branches and the like. Batteries can heat up when being charged at high rates, such as would happen at high noon on a summer day, but again, ventilation, or air conditioning, can keep that from becoming a problem. Inverters will usually have an internal fan that comes on at a given temperature. If the ambient air temperature is kept cool enough, they will cool down readily. So depending on where you are located, you might expect to have to condition the air surrounding the electronics. In Phase One, we simply left a crack between a 2 x 4 and the Solar Panel-holding OSB panel. Since water would have to run uphill to get inside the battery compartment, we thought that would be safe, but water has been able to get in, apparently when there is a breeze blowing up the face of the OSB panel that holds the solar panels. There is also an electronic box on the back of each solar panel, but one of our suppliers has told us that there is no problem with allowing that box to contact the roof, or not allowing any ventilation space underneath the solar panel's frame. All in all, you should get the standard lifetime out of these devices as long as you keep them cool and dry.

Appendix

Sources:

SOLAR PANELS:

NewPowa America Corp: www.newpowa.com
Missouri Wind and Solar: www.mwands.com

These companies are both E-Bay vendors, so if you prefer to deal through E-Bay, you may do so, but you may, in fact, be able to get better pricing by buying direct. You can still pay through PayPal if you prefer, regardless of how you contact the company.

CONTROLLERS:

Both companies listed above also have controllers and inverters, but you can find lower-priced controllers and inverters at these other vendors:

Frentaly, a 5-star retailer on E-Bay since 2004 (also on Amazon)
https://www.ebay.com/usr/frentaly?_trksid=p2047675.l2559
This is a very low-cost PWM controller, not MPPT

Another PWM Controller, this one claims to be waterproof:

https://www.ebay.com/itm/10-20-30-40-50A-Waterproof-Solar-Panel-Charge-Regulator-12-24V-LCD-Controller/112580580000?

This is the MPPT Controller we have been using (Must have red stripe across the front of the unit):

https://www.ebay.com/itm/30A-12V-24V-Auto-Switch-MPPT-Solar-Panel-Battery-Regulator-Charge-Controller-US/273050205573?

This one looks good (more than twice the price, however):

https://www.ebay.com/itm/Effective-Solar-MPPT-T20A-12V-battery-Charge-Controller-Extension-cable-US-WT/192429642902?

This looks to be very good, heavier construction, more money (This one is only ten amps current, requires 24-volt minimum system):

https://www.ebay.com/itm/Docooler-MPPT-Solar-Panel-Battery-Regulator-Charge-Controller-10A/282841153803?

Similar product at lower cost, ships from China

https://www.ebay.com/itm/LCD-MPPT-Solar-Regulator-Charge-Controller-24-36-48-60-72V-Boost-MPT-7210A/112451914404?

INVERTERS:

Dr. Powerjack (this is new 2018 version of the 8000-watt unit we bought). Note that this can also be used as a charger to quickly charge your battery pack, either by plugging it in to a power line, or by using a wind generator or other source of electricity:

https://www.ebay.com/itm/8000W-LF-Split-Phase-PSW-power-inverter-DC12V-AC110V-220V-Charger-UPS-ATS-2018-V/253273613491?

This is a much smaller similar item, not really suited for whole-home use:

https://www.ebay.com/itm/3000W-LF-Split-Phase-PSW-12V-DC-110V-220V-AC-60Hz-Pure-Sine-Wave-Power-Inverter/263386755523?

Here is a decent 110-volt inverter, good for a much smaller system, uses 24-volt input instead of 12-volt:

https://www.ebay.com/itm/Power-Inverter-1500W-24V-to-120V-Pure-Sine-Wave-Inverter-US-Plug-Solar-System/191734723230?

This is a 6000-watt AIMS inverter, much more expensive, appears to be true split phase with the neutral wire grounded. Note that it requires 24 volt input:

https://www.ebay.com/itm/AIMS-Power-6000-Watt-24V-DC-to-120-240V-AC-Split-Phase-Pure-Sine-Inverter-Charge/263279971120?

Should any of the above items not be exactly what you need, search for a similar item that has the other features/functions that you need.

APPS WE LIKE: (These are for Android Devices, but the same or similar apps should be available for Windows and IOS devices.)

Electrical Calculations:

This one is for DC circuits only:

Circuit Wizard
Blue Sea Systems   Tools
E Everyone
★ ★ ★ ★ 201
① This app is compatible with all of your devices.

Electrical Calculator
Navaile   Productivity
E Everyone
★ ★ ★ ★ 1,303
① This app is compatible with all of your devices.

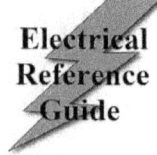
Electrical Reference Guide
Premier App Designs   Tools
E Everyone
★ ★ ★ ★ 14
① This app is compatible with all of your devices.

Add to Wishlist   Install

Learn Electrical Wiring
Muzakkir   Books & Reference
E Everyone
★ ★ ★ ★ 247
Contains ads
① This app is compatible with all of your devices.

Home Electrical Wiring Diagram
oneapps.edu   Books & Reference
E Everyone
★ ★ ★ ★ ★ 13
Contains ads
① This app is compatible with all of your devices.

Measuring Devices

SOLAR ENERGY On A Shoestring Budget

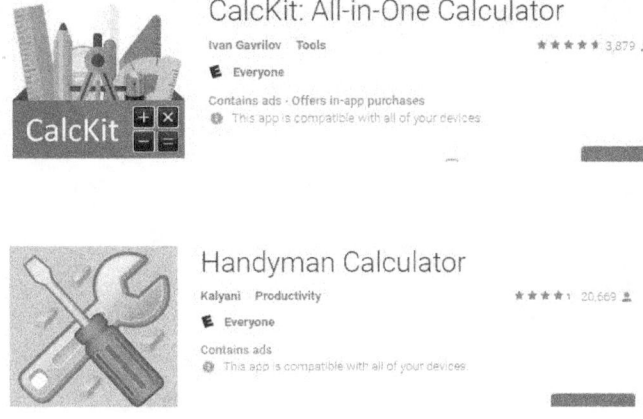

Inclinometers and levels:

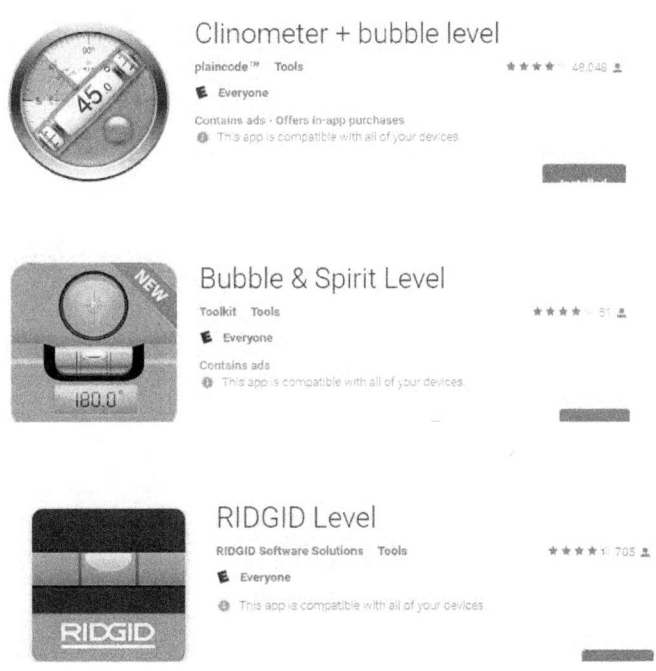

Rafter Calculators:

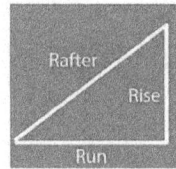

Rafter Length Calculator

David King LearnFraming.com Tools ★★★★ 7

E Everyone

Contains ads
ⓘ This app is compatible with all of your devices.

Rafter Calculator

Armogo, Inc. Tools ★★★★ 95

E Everyone

ⓘ This app is compatible with all of your devices.

SOLAR ENERGY On A Shoestring Budget

These are some photos of Phase One of our system in the early construction stages.

The first picture directly above shows how we hung the solar panels; a small block of wood (1" x 4" x about 8" long) was screwed into each corner after we drilled a hole in the frame for the screw. Then the panel was laid in place and screwed to the roof from the underside of the roof. Be CERTAIN that your screws are not long enough to go into the back of the solar panel, which is vacuum sealed! (That's the white area under the panel, inside the frame. That may not be compromised!) You should use screws that are flat on the bottom of the head. If you can't find wood screws of this type, sheet metal screws can be used. You also could use a washer under the head of the screws. But you definitely do not want to damage the backing of the solar panel. The second photo shows the extension we built to hold three more solar panels, and provide some weather shielding from light rain and heavy sun. Then we added a voltmeter so we could be aware of the battery condition without having to open the drop-down door, which is visible in the center photo, and is secured with a piano hinge at the top, and a drawer lock in the hole at the bottom. We used two screw-in brackets that we can just flip, to hold it open. You could use an automotive lift gate gas strut, which would be handier....

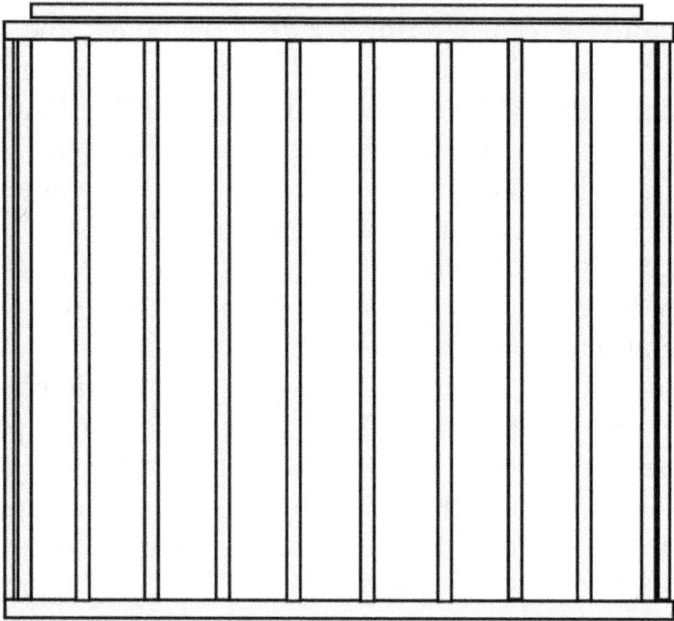

The tall and short walls of the shed would be built like this, with studs 16" apart, double studs on the ends, a single footer across the bottom, and two headers across the top, with the top one cut short so that the top headers from the side walls can be placed over them after the side walls are in place. This ties the four walls together in a rigid rectangle. Because the walls are built on the floor, the headers and footer can be fastened from the outside, with nails or screws going straight into the ends of the studs. This is much stronger than toe-nailing them. The tall wall should be built first, then raised into place, then the short wall can be built and raised into place. They would temporarily be held vertical with spare studs fastened from the middle of the outside stud to the middle of the outside floor joist. This should be done at both ends. Then when you build the end walls, they can be raised, and the end studs on the end walls can be fastened to the double studs on the tall and short walls.

Two ways to build the side wall headers

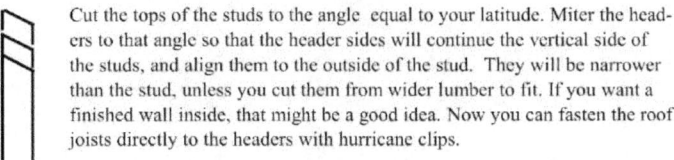

Cut the tops of the studs to the angle equal to your latitude. Miter the headers to that angle so that the header sides will continue the vertical side of the studs, and align them to the outside of the stud. They will be narrower than the stud, unless you cut them from wider lumber to fit. If you want a finished wall inside, that might be a good idea. Now you can fasten the roof joists directly to the headers with hurricane clips.

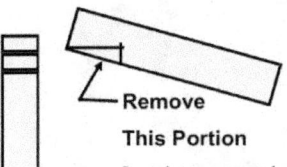

In order to use studs with horizontal tops and horizontal headers, you must fly-cut the joists so that they fit solidly to the horizontal tops of the headers, while the joist itself is at the proper roof angle. You might also cut the end of the joist so that it is vertical with the joist itself at the proper roof angle. The joist header should create a drip line that is outside of the outer wall, unless you want to let the roof overhang the joist headers, in which case the roof itself will create the drip line outside of the outer wall.

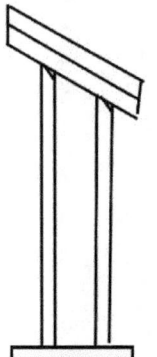

The studs on the end wall will decrease in height with every stud, to match the angle of the roof. Those studs should also be mitered on the top, to match the angle desired. The headers will then lay nicely on top of the studs, provided you cut the studs to the proper length.

The decrease will amount to the difference in height over a 16" horizontal length. You can calculate this by using the formulas given earlier, and using a base of 16". You should measure all of the studs at the high point. In this diagram, you are looking at the short side of the stud—the 2" side of a 2 x 4 stud. The bottoms will stay at 90 degrees. Another way to look at this is that the triangle between the top of one stud and the top of the next stud, and the point where a horizontal line from the top of the shorter stud meets the tall side of the next taller stud, is exactly the same proportions as the triangle that defines the roof.

This photo shows how a building is tied down using a screw auger. The device is buried into a hole in the ground, with the eye at the top lined up with the floor joist headers, and held to the building through the end of a joist with a Lag Screw and a heavy washer. Caulking prevents any water from going around the screw, and through the washer to the wood itself. This was a commercially-built shed used as a motorcycle workshop elsewhere on John's property.

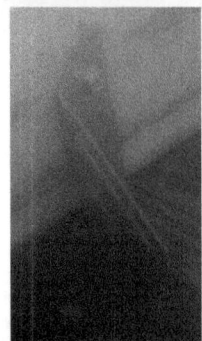

These are some photos of a similar 10' x 12' deck built to hold an Arrow steel shed. The first shot shows the completed frame without the plywood deck. The second shot shows how the joists were held to the PT timbers with hurricane clips. The third shot shows the blocks in place and leveled, with one timber in place, and the other one ready to go. The blocks on the right have been leveled with sand under them.

About the Author:

Jay Hamilton was a freelance photojournalist specializing in motorsports, through most of the 1970s. His work was published in national magazines such as "Cycle Guide" and "Modern Cycle" as well as smaller publications. He had his own darkroom where he processed black and white as well as color photographs, and color slides as well. He still enjoys riding his motorcycles, though he drives a mini van these days, instead of a collection of Bug Eye Sprites that he no longer owns. His collection now is limited to a Chrysler Town and Country, a Honda Shadow 750, a Honda CL-360 from the 1970s and a 1982 Honda Civic Wagovan. He uses digital cameras these days, and would love to have one of those new Lumia 950 phones with the 20 megapixel camera built in. He is a bit of a computer nerd, but does not have the funds to own any real mega computers, though he does have a Galaxy Tab S 8.4" high def tablet, which fits in his pocket and largely takes the place of his Gateway laptop, and an old P4 desktop that he built from scratch. He likes dogs, and very much loved his Chocolate Lab puppy that was dognapped from his home one Sunday morning a couple of years ago.

He is looking for a replacement. He also has several Calico cats. He is working on a solar installation to reduce electric costs, and has a Champion 3500 watt generator wired into his power pole through a breaker, so he can manually switch between the power utility and his own power. He is an avid gardener as well. Like the rest of us at Crossroads Publishing, he hopes the sale of his books will supplement the meager check he gets from Social Security (being a freelance photojournalist is not conducive to having a large Social Security check.) He is a hopeless liberal, not at all happy with Republicans, but not fond of Obama either. He fully expects to live well past the age of 90, and is in very good health, and doesn't look anywhere near his real age, which he refuses to disclose.... And yes, he is still single, and expects to remain so.

Other books by Crossroads Publishing of Florida

We have a website up and running at **www.cpubfl.com**, where we have several other books as well. Flip the page to see a few of them, and you can check our website for the most up-to-date list. For links to purchase these books, please go to the landing page for our website at **http://www.cpubfl.com**. We will be putting many more of these books on other websites, both retailers and aggregators. Links on the website will be updated as we do that. In particular, Jay's new book EXERTING INFLUENCE, which is a political primer for millennials not happy with the 2016 election, provides a roadmap for repairing the system for 2018 and 2020, and for influencing the reader's representatives and senators in the meantime. There is an appendix in the back which shows where to find your representatives and senators in every state, and where to find the full text of all of the laws in every state. This should be required reading for every new voter, especially, of course, those inclined to be progressive.

And PLEASE—if you have found the information in this book to be of any value at all—leave a favorable review on the website where you bought the book. Factual reviews are important to a struggling new writer or publishing company. It takes fifty reviews before book distributors will seriously help promote a new book, so this is of the utmost importance to us, to help us keep bringing you more high-quality books like this one.

Check out these other fine books from our friends at Crossroads Publishing of Florida, at www.cpubfl.com

This is a book for Bernie Sanders fans who are upset about election 2016, and want to do something about it. The book offers an introduction to how our political system works, how to influence voters and legislators, how to find out who to contact, and how to contact them. There is a full appendix with links to the principal websites in each state to look up legislative information and legal statutes.

See ordering details at www.cpubfl.com

When you walk into a car dealership to look for a car, who do you think has the advantage? Does the customer have the advantage, since he has the money? Not on your life! Even if you buy a new car every two or three years, the salesperson who is about to greet you talks to three or four customers EVERY DAY! The dealer has this down to a science, and you don't. This book will tell you what to watch out for, and how the salesperson is trained to control the customer.

See ordering details at www.cpubfl.com

These are just some of the books you will find on our website at www.cpubfl.weebly.com

We are three old names in photojournalism from the seventies, who have come together in Crossroads Publishing of Florida, as authors.

John Waaser was the Eastern US Contributing Editor of "CYCLE WORLD" in the mid-to-late 1970's. Jay Hamilton covered some of the same events for "CYCLE GUIDE" and "MODERN CYCLE," while Glenn Stewart sold photos, mostly to "CYCLE" Magazine. Now retired, they have combined their talents to produce e-books and paperbacks.

BOOKS BY JAY HAMILTON

When you walk into a car dealership to look for a car, who do you think has the advantage? Does the customer have the advantage, since he has the money? Not on your life! Even if you buy a new car every two or three years, the salesperson who is about to greet you talks to at least three or four customers EVERY DAY! The dealer has this down to a science, and you don't. This book will tell you what to watch out for, and how the salesperson is trained to control the customer.

See your retailer or ordering details at www.cpubfl.weebly.com

The Entrepreneurial Mindset
GROW YOUR BUSINESS INCREASE SALES, and EXPAND YOUR BRAND!

Jay Hamilton

This book puts into clear and simple terms, how to develop the mindset necessary to become a successful Entrepreneur. The ten chapters go into habits, tips and tricks, technology, including some apps that we like and a few words on which is better, Android or Windows, for your mobile device. Our conclusion might just surprise you! And we end with a few mental tricks to remember. There are things in here that you will need to know, and there are things in here that you might be surprised either that you did already know them.

See your retailer or ordering details at www.cpubfl.weebly.com

BOOKS BY GLENN STEWART

The Small business owner has to market smartly in order to grow. And to do so, he must build a network of people around him who will assist him in every way possible. This book describes this process in clearly understandable terms and with checklists at the end of each chapter.

See Your Dealer or see Ordering Details at www.cpubfl.weebly.com

This is a book that covers the basics of choosing your business, studying the niche, jumpstarting some quick income, building on your success, and finally quitting your steady job to run your business full time.

See Your Dealer or see Ordering Details at www.cpubfl.weebly.com

BOOKS BY JOHN WAASER

This book is full of ways to improve your memory and recall ability, including diet, supplements, and exercise. Great for older people, and people who just can't seem to remember things readily. We talk about foods and eating changes to boost memory. We go into herbal supplements, and Aromatherapy, and then we go into brain games. Memory prompts and mnemonics can also be used to help bring things out of your inner mind.

See your Dealer or see Ordering details at www.cpubfl.weebly.com

A look at chronic fatigue, and how you can fight it. Natural habits that you can stack up to help are discussed, along with Teas that will pick you up quickly, foods and supplements that can help, and essential oils and other tips. This is a complete book on fatigue and fighting it.

See your Dealer or see Ordering details at www.cpubfl.weebly.com

BONUS READING!

From How to Buy a Car Without Losing Your Shirt

SOLAR ENERGY On A Shoestring Budget

TABLE OF CONTENTS

Cover Art

About The Author

Copyright Information

Chapter One: General Information

Chapter Two: The Five Steps in Selling a Car

    1: Greet the customer

    2: Walk around the car, pointing out features

    3: Get the customer to DRIVE the car

    4: Negotiate and close

    5: Delivery

Chapter Three: Buy From the Invoice

Chapter Four: The Quota System

Chapter Five: The Trade-in Allowance

Chapter Six: Financing Your Purchase

Chapter Seven: Buying a Used Car

Chapter Eight: The Service Department

Chapter Nine: Warranty Work

You Are Now Ready To Buy A Car

CHAPTER 1:

GENERAL INFORMATION

When you walk into a car dealership to look at a new car, who do you think has the advantage? Does the customer have the advantage, since he has the money? Not on your life! Even if you buy a new car every two or three years, the salesman who is about to greet you talks to three or four customers EVERY DAY! He has this down to a science and you don't. This booklet cannot make you a wonder buyer. The thing you must do is to be firm, and remember everything in this booklet, and that will be difficult. It would not hurt a bit to practice, with someone else playing the role of the salesman. (Role-playing is one of the first things a new car salesman does, before they turn him loose on the sales floor.) It might not even hurt a bit to "buy" a few cars from other dealers before you go to the dealer you want to buy from. Repeat players have all of the advantages, and you want to turn yourself into a repeat player before you play for real.

Choosing a dealer may be even more important than choosing the car, and the chances are you could live with something other than your dream model car if the dealership were more attentive. You should ask all of your friends who have bought cars lately what they think of the dealers they bought from.

You should find out if there is a high turnover of personnel at the dealership. That is a common problem in the automotive industry. If the salesman leaves the dealership a month or two after you buy a new car, you have lost a lot of leverage. The salesman will ask you to send him any friends who are about to buy a car—and he means to send them to him personally, not just to the dealership. This is because he is paid a commission on every car he sells. That commission is not like the commission a salesman in a retail store might get, of about 5% of the purchase price. The car salesman gets a commission of 25 – 30 per cent of the Gross Profit on the deal. The more money he makes for the dealership, the more he puts in his pocket—and he gets a considerable chunk of the difference. If you send him a few customers, he earns a lot of money by keeping you happy. Don't be shy about asking him for a referral fee, known in the trade as a "bird dog" fee. These days that fee is likely to be $100 per closed deal. It certainly should be no less than $50. It is best if you take your friend in there and personally introduce him, or at least call the salesman BEFORE the friend gets there, so there is no doubt that the reason this customer saw that salesman is because you sent him in. For your information, the bird dog fee is generally paid by the dealer and does not come out of the sales rep's commission. This kind of potential relationship is enough to keep the salesman who sold you the car, interested in any problems you have with the car after you get it home. If he leaves the dealership, you have a lot less leverage in case of problems.

For this reason, I do not recommend buying a car over the phone, or over the internet. I have seen books which advise you to talk only to the "internet salesperson." Poppycock! In dealerships where I have worked, every salesperson was entitled to sell over the internet, and from internet offerings I have seen, that appears to be a common thread. But if there were an internet-only salesperson, they would most likely not be on commission, they would be a low-salaried employee, with little knowledge about cars and/or the workings of the dealership. They would be simply an order-taker, not capable of helping you choose a model that is right for you or anything else. Also, like dealerships which advertise that every vehicle is pre-priced, and their sales people are not on commission, the price includes an average markup, which gives the dealer the kind of profit he expects to make on every deal, and you want to hold him to less profit than that. If you read this book through, and understand the process, and get some practice before you negotiate for real, you should have no problem in working the dealer down to a much lower profit than those single-priced dealerships, or internet salespeople, will be permitted to offer you.

It goes without saying, that because of the way the salesman gets paid, you should always talk to the same salesman every time you return to the dealer. It has become apparent to me that many customers do not realize this. But that salesman will get to know you, and—as we already noted—he will be your strongest ally in the dealership, should a problem develop. Suppose you talk to one salesman a couple of times, and then buy from someone else: the first salesman has essentially wasted every minute he spent with you. If you come back with a problem, he will NOT want to talk to you, PERIOD! If, on the other hand, you always talked to the same salesman you bought from, and he is not there when you come in with a problem, someone else may be willing to help with your problem, because they would want someone to help THEIR customer under similar circumstances. Solving a problem after the sale is not a waste of time. It is a duty they owe to the dealership. But talking to someone who later buys from someone else IS a waste of time, and a salesman who has wasted time with you before the purchase, will not want to know you after the purchase. Other salesmen may also be aware of your fickleness.

If, on the other hand, you don't like the salesman with whom you are talking, you should ask to see the sales manager, and explain this to him as soon as you become aware of your feelings. Every salesperson has their own style, and most dealerships will keep people from a variety of backgrounds there. It is far from unheard of that a particular customer will not like one particular salesperson from time to time. But generally, there will be some hard feelings somewhere, if one salesperson wastes a significant amount of time with a customer who then buys from another salesperson at the same dealership. The customer may wind up being one of the ones hurt by those hard feelings, so please, Please, PLEASE always talk to the same salesperson whenever you go into a particular dealership. If you don't see him, ASK for that person by name. You both will generally be better off that way. Normally, if another salesperson writes a deal after you have asked for your salesperson, the two will split the commission. It would be best if you CALL your salesperson, and ensure that (s)he will be there when you plan to arrive. Unless the shop is unusually busy, (s)he will try to avoid talking to anyone else within an hour or so of any appointment. Of course, if (s)he has left the dealership, you might want to know why, and where (s)he has gone to (something the dealership might be reluctant to tell you, since it could cost them business.)

Once, I had to leave to go to a locksmith to pick up a key, when we could not find the key for something I was selling. When I arrived back, I saw a regular, steady customer of mine in the showroom. I waved to him; he lit up in a huge grin, and waved back. As I walked out of the closing room again for a second, I told him that I would be right with him. I then noticed on the computer that another salesman was working with him. This other salesman was (in my well-considered opinion) a pathological liar. I immediately walked up to my customer, and asked him whether he had asked for me when he first came in. He replied that he had wanted to talk to me, that he had asked for me, and that he had been told simply that I was not there. The other salesman absolutely insisted that that the customer did not ask for any salesman when he first walked up to the customer. I have waited on that customer practically every time he came in to the shop, even for parts. I had the kind of relationship where I am certain that the customer would not walk into the shop without my name on his lips. He was looking at a particular model, which he had discussed with me earlier. We had the 1993 model, which had come in since he had last talked to me, in the back under wraps until we sold the 1992 unit that we had in stock. I know that this customer wanted the '92 over the '91 when he bought his trade. I certainly would have told him that we had the '93 in stock. Instead, the other salesman made him a deal on the '92, without telling him that the '93 was in. Ultimately the deal fell through, probably because the customer sensed that the other sales rep was lying about whether the customer had asked for me, since the other salesman wanted to keep the deal for himself, rather than splitting it with me. The commission on the deal was approximately $160, so the salesman was lying for about $80. The boss will not call the customer to verify the truth. I probably will ultimately quit, and if I do, this incident will have a lot to do with it. (Actually, I did quit, and started my own computer store, a short while later, after the same salesman pulled similar stunts on both myself and the third salesman in the shop. Once he even told a customer that the third salesman was gone, "but he'll be back in a (sic) hour…" The other salesman had gone home to watch a major race on television, and had said he might be back before closing, but he was having friends over, and there were sure to be martinis involved, and nobody expected to see him return. When the customer came back, this guy started a deal in the computer, telling the customer that it was just roughed in, and that his regular salesman would finalize things like the trade allowance. This customer, a retired professor, was buying four or five vehicles a year from one particular salesman in this dealership. Store rule was that the salesperson who started the deal in the computer owned the sale, and that if it was ultimately closed by another salesperson, the commission would be split. I couldn't wait to get out of there.) If you find yourself in a similar situation, and you perceive that one salesman has lied to you or about you, do everyone a favor, and speak to the owner of the business, and tell him WHY you will not do business there, and why you will not recommend that your friends do business there. Of course, if you still like your regular salesperson, you can calm down later and continue to do business with your regular salesperson….

The "Service Advisor" who greets you when you come in for service is also paid a commission on his sales, and it is therefore in his interest to sell you service that may not be necessary. If he knows that in the event he treats you properly, you will always come back to him, that represents repeat business which may in the end be more lucrative than stabbing you in the back. Again, if there is a high turnover of service writers, you will lose that leverage.

Advertising is viewed by car dealers as simply a way to get customers to the lot. The sales process does not begin until you get to see a salesperson. Dealers will flat lie in their advertising, and the salesman is taught to lie to you if you telephone in from an ad. They will say that a car, which is already sold, is still available, for instance, and if you put down the phone and race to the lot, they will say that another salesman just sold it after you called about it. Many dealers now advertise a price that is $2000 below the actual price, but there is fine print, which says "with $2000 cash down or trade equity." The price is so much lower than other dealers are advertising the same car for, that you are bound to notice, and to visit this dealership—probably first on your list. That is the whole purpose of advertising. Remember that this is a numbers game. For every four people he gets on the lot, the dealer will sell about one car right away. He can expect to sell at least one of the others later, if his salespeople do their follow-up correctly. His primary concern with advertising is to get you on the lot. He doesn't care if his advertising is honest. The courts will allow a little "puffing" because you are expected to notice the puffing. Caveat Emptor (Buyer Beware) is still the basis for the legal system in this country, even in those states with strong consumer protection laws. Most of those laws don't really have the teeth the legislators thought they would have when they passed the laws, and they certainly don't have the teeth the consumers thought they would have when the news outlets ran stories on these laws. DO NOT TRUST A DEALER'S ADVERTISING!

The basis of the legal system in this country, when it comes to commerce, is something called the Uniform Commercial Code, or at least that is supposed to be the basis. Part of that code is the concept that there is an implied warranty of merchantability, and of fitness for the intended purpose. I recently discovered that there are ten or a dozen states in this country where all the dealer must do is to include the words "As Is" on the bill of sale, and in those states (Florida is one of them), those four letters and one space, wipe out the implied warranties guaranteed by the Uniform Commercial Code. This is a rip-off, and occurs in states where the automobile dealership lobby owns the state legislature, and this is true even in states which try to tell you that they have strong consumer protection laws (like Florida.) I would hazard a guess that most of these states are traditional Republican strongholds. Do a search using your favorite search engine, for "implied warranties in (your state)." There should be some consumer-oriented law firms in the state with websites dealing with this subject, and they should come right up when you enter that search string. This applies predominately to used car sales, since new cars are warrantied nationwide by the manufacturer.

We said that the dealership has the art of selling you a car down to a science. They keep records so detailed that they can tell which salesperson has the best close ratio for first visit customers, and which salesperson works the hardest to get unsold customers back into the dealership, and so on. There are also several STEPS in dealing with each customer that are practiced religiously by all car dealers. These steps may vary slightly because the dealer subscribes to a sales organization that tells him how to best turn a prospect into a sold unit, and those organizations vary slightly in what they tell the dealers to do.

Those sales organizations also break down their dealers into groups of ten to twenty dealerships, usually based on demographics of the dealer's locale. These might be called a "Twenty Club," or whatever. One of the dealers I worked for belonged to a sales group that I called "The Lemmings." Oddly enough, one of the other dealers in his 20 club was a guy I had known personally, a couple of decades earlier. Each month, usually on, say, the third Thursday of each month, but generally the same day of the week, and about the same time of the month, all twenty of them will descend on one of their dealerships, and break down what the dealer is doing right, or perhaps wrong. If this dealer has found something that works well, he will show it off, and the others may copy him. If someone spots something he's doing wrong, they will bring it to his attention, and suggest ways to improve the situation. Often, the companies who own these businesses that show dealers how to run their dealerships, are guys who have lost one or more of their own dealerships!

Often, they hire a psychiatrist to script their sales routine. In some cases, sales people will be ordered to follow the script word for word, while in other cases, they may have more freedom to divert, as long as they cover the bases. But there will be key words that the psychiatrist has assured them will break down resistance. It would be interesting to record an audio transcript of the entire sales pitch, to see if other dealers will be using the same script. I will try to get a blog up shortly after this book is published so that people from all over the country can upload such recordings for comparison, or upload notes with specific quotes that seemed a bit out of line, so that other buyers can listen for early clues that their dealers are about to unleash something dishonest on them.

(The real nitty gritty is in chapter two and beyond....)

And now, more bonus reading from John Waaser's book, PHOTOGRAPHIC COMPOSITION

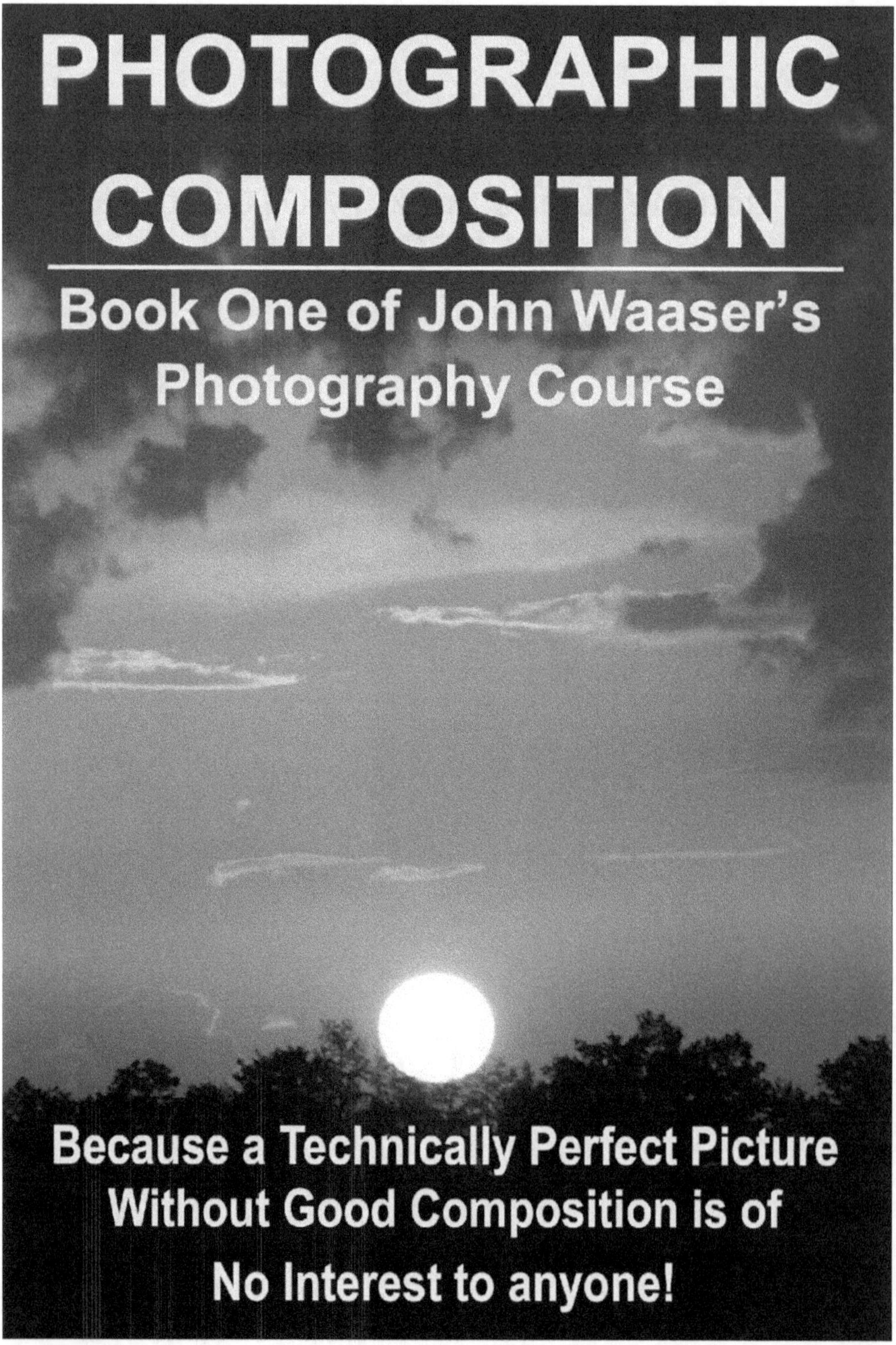

PHOTOGRAPHIC COMPOSITION

Book One of John Waaser's Photography Course

Because a Technically Perfect Picture Without Good Composition is of No Interest to anyone!

SOLAR ENERGY On A Shoestring Budget

PHOTOGRAPHIC COMPOSITION

Book One of John Waaser's Photography Course

Because a Technically Perfect Picture Without

Good Composition is of no Interest to Anyone!

By John Waaser

© 2017

This book copyright 2017 by

Crossroads Publishing of Florida

P. O. Box 222

Worthington Springs, FL 32697

www.cpubfl.com

All rights reserved

This document is geared toward providing exact and reliable information regarding the topic and issue covered. The publication is sold with the idea that the publisher is not required to render legal, accounting, officially permitted, or otherwise qualified services. If advice is necessary, legal or professional, a practiced individual in the profession should be ordered.

- From a Declaration of Principles which was accepted and approved equally by a Committee of the American Bar Association and a Committee of Publishers and Associations.

In no way is it legal to reproduce, duplicate, or transmit any part of this document in either electronic means or in printed format. Recording of this publication is strictly prohibited and any storage of this document is not allowed unless with written permission from the publisher. All rights reserved.

The information provided herein is consistent, in that any liability, in terms of inattention or otherwise, by any usage or abuse of any policies, processes, or directions contained within is the solitary and utter responsibility of the recipient reader. Under no circumstances will any legal responsibility or blame be held against the publisher or author for any reparation, damages, or monetary loss due to the information herein, either directly or indirectly Respective authors own all copyrights not held by the publisher.

The information herein is offered for informational purposes solely, and is universal as so. The presentation of the information is without contract or any type of guarantee assurance.

The trademarks that are used are without any consent, and the publication of the trademark is without permission or backing by the trademark owner. All trademarks and brands within this book are for clarifying purposes only and are the owned by the owners themselves, not affiliated with this document.

Introduction:

John Waaser was the perfect person to write this book. As a student, both at Mount Hermon School for Boys and at Northeastern University, where he studied Mechanical Engineering, he was elected Vice-President of the camera club for two years running, four years total, where his principal duties consisted of coming up with the topic for each meeting, and securing the educational material for that topic from top companies such as Eastman Kodak and Ansco. It also fell to him to find someone to proctor the discussion, which about half of the time, he did himself. He later became a freelance photojournalist, and for two decades, he principally photographed motorcycle races, and other motorsports-related activities, including an occasional road test and other features. As a journalist, he showed an ability to take a highly technical subject and break it down so that ordinary people without a technical background could understand it. He took any number of portraits of up-and-coming racers as well. He did a few portfolios for models, and he photographed a few weddings. He spent about a year as assistant editor of a biweekly tabloid newspaper, where he wrote copy, took photos, set advertising, and laid out the pages. He owned Adpho Graphics, a photo studio and advertising agency, in the early 1970s. His personal hobby has long been night-time available light photography outdoors, where he frequently hand-held exposures of up to 30 seconds. He had his own photo lab at one point, where he processed film and prints including both black-and-white, and color negatives, and color transparencies (slides) as well. He constructed a film dryer and an enlarger stand with variable height easel shelf, and published articles and photos of their construction in "Popular Photography" Magazine. For several years, he was listed on the masthead of "Cycle World" Magazine as their Eastern US Contributing Editor. He has owned a computer store, and has owned digital cameras since they had VGA resolution or less. He now owns an Olympus E-PL1 camera with two lenses, and carries several phones and/or tablets at all times. He also taught an adult education second-year photography course at a local community college for two semesters, while one of their regular professors was on a sabbatical.

John remembers the time he met the playwright Terrence McNally. Terrence had just produced his first play, then, at the Yale School of Drama. He had a crippled dog named Charlie, who he adored. John took a portrait of Charlie that Mr McNally said captured Charlie's essence better than any photo he had seen. John sent him several prints of that photograph, one of which was a large (16 x 20 or 20 x 24-inch) print that was Sepia Toned. Sepia toning was a process for black and white photographs that converted the silver in the image, to a compound that was much less sensitive to light. Several years later, one of the photography magazines stated that Terrence had become an avid amateur photographer. John has often wondered if his photograph of Charlie drew Mr McNally into photography as an avocation. And since the photo is probably still in good condition, John frequently thinks about whether it still graces a place of honor in his home. And he would like to think that you might become an avid photographer after reading this book.

SOLAR ENERGY On A Shoestring Budget

Table of Contents

Chapter 1: General Discussion

Chapter 2: What is the relative size of the subject?

Chapter 3: Location for the shoot

Chapter 4: Subject location within the photo

Chapter 5: Lighting

Chapter 6: Background

Chapter 7: Foreground

Chapter 8: Viewpoint

Chapter 9: Color

Chapter 10: Special Considerations for Portraits

Chapter 11: Fireworks!

Chapter 12: The Eclipse

Conclusion

Bonus Reading

Copyright information

Introduction

Other Books from Crossroads Publishing of Florida

Chapter 1: General Discussion

I felt it important to discuss composition before even telling you how to operate your camera. For one thing, the owner's manual for the camera has more specific information in that regard, and you should consult that and practice using the camera until you are quite used to the functions of that particular camera. But the biggest reason why I thought composition was the most important first lesson, is that a technically perfect photo that lacks good composition is of no interest to anyone. I have sold photos for good money, that were lacking in technical perfection. I once sent a 120 roll-film transparency to "Cycle" Magazine, without magnifying it to see if it was perfectly sharp. They loved that photo, and ran it as a full-page inside color photo. When I saw it in the magazine, I was shocked. It was blurred very badly, possibly due to focus, maybe camera shake, or a combination of factors. I was absolutely ashamed to see that photo in print, yet they probably gave me about $300 for it (they paid ASMP—American Society of Magazine Photographers—rates, which were quite high, much higher than I was used to getting.) I questioned why that photo was run as a full-page photo, and they said that they loved the fact that three riders going around a tight 180-degree downhill curve at Bryar Motorsports Park in Loudon, New Hampshire, looking like they were stacked one on top of another, was unusual, and they just loved the photo, and thought it was deserving of full-page status in spite of its technical faults. There could be all kinds of reasons for technical problems, but when a shot is one-of-a-kind in terms of composition, and all of the elements come together perfectly, technical perfection is not an issue. People will love it. That is why composition is the first lesson in this photography course. Learn to take good-looking photos first, then learn the technical minutia.

Another similar story of a photo that was marred by technical details, but caught the moment perfectly, and the family bought numerous copies of the photo, was when I walked out of a pit garage at Daytona International Raceway, and spotted the son of a former world champion racer, dumping his mini-bike in the pits. I whipped the camera up to my face, and took the picture. I had just been using available light for close-up pictures of an engine being worked on, so the aperture was fairly wide open, and the shutter speed very slow, and the lens was focused for just a few feet. The resulting photo was overexposed and out of focus, and very blurred. But it caught their son just as he hit the ground, and they loved it! Technical faults mean nothing at a time like that! What matters is that I got a picture that nobody else got. The blur served to show that he hit the ground while moving, so it may actually have enhanced the photograph.

And composition covers any number of areas. The subject and how it is posed, is one big one, but the foreground and background, the lighting, and the color balance all enter into it. What leads the eye to the subject? What separates the subject from the background and the foreground? Is the lighting good? Is the subject at the best angle? Do shadows, or blocked up highlights, detract from the overall view of the subject? Is there a bunch of unimportant flotsam and jetsam around the subject to detract from it? The perfect photo must be carefully staged. A few small things might not detract too much, even though some will notice them. Case in point: the cover photo of this book. I should have stepped a little to the right, to center the sun into the dip in the tree line at that point. I did not notice that in the screen on the back of the camera while shooting, but I sure did notice it in the finished picture, when I blew it up. But that wasn't a really major problem, and I was able to select the sun color to brush over the few twigs that were in front of the sun, and you would never know that they were there. I used a free app (well, I paid for the pro version, but you can certainly start with the free one) on my Android tablet to do that. The app I used was PhotoSuite by Mobi Systems, who also make OfficeSuite and File Commander, both of which I have loaded in all of my Android devices. And I cropped the picture (using the tools in Microsoft Publisher, a free download when you run Microsoft Office 365), and made some other changes that brought the sun to the center of the photo, even though the subject should ideally be one-third of the distance from the two closest edges for a balanced image. But I was more concerned with the cloud formation at the top, and I wanted it to frame the photograph, coming down the sides the same amount on either side. So I lived with the sun in the center. I believe that this photo still is in perfect balance because of the way I cropped it, even if the right-side cloud is somewhat heavier. Question: what is the subject here? You can have a sunset picture that doesn't even show the sun. The clouds are generally the subject in a sunset picture. Here, I would argue, the sun is the subject. So where is the subject focused? The subject is staring straight at you, the viewer. Therefore, I believe that the center is the one position which will truly give balance to this photo. The object that the subject is focused on is, in fact, the viewer, who is outside of the photo, and presumably front and center. These are choices you sometimes have to make. I had dozens of photos of that sunset, and this was the most appealing one, to me. So I went with it, and lived with the minor faults that existed or that I created, in cropping it. It's also closer to the bottom than one-third up, but it's close enough, and I wanted enough dark cloud up at the top to permit me to use white lettering. I had plenty of dark area at the bottom. I have had numerous comments about how much people liked the photo on the cover. I have noticed that when shooting sunsets, where you point the camera is important, in determining the exposure. Your camera will have a way to set the exposure from the meter, to give you the best image, and then move the camera to get more foreground or background into the image. Usually this involves simply holding the shutter release partway down, short of actually tripping the shutter.

Composition is composed of many areas of concern. There is the position of the subject, the position within the photo itself, whether the subject is leading you to something else within the photo, or to someplace outside of the photo, lighting, shadows, color values, and a number of other things, which we will go into shortly. The more technical ends of photography will be saved for future books in the series.

www.ingramcontent.com/pod-product-compliance
Lightning Source LLC
Chambersburg PA
CBHW070157230526
45471CB00002B/697